D1558979

# The Evolution of Clastic Sedimentology

Coloured geological map of England, Wales and part of Scotland by William Smith (1815).

# The Evolution of Clastic Sedimentology

Hakuyu Okada

*with*

Alec Kenyon-Smith

DUNEDIN ACADEMIC PRESS
*Edinburgh*

Published by
Dunedin Academic press Ltd
Hudson House
8 Albany Street
Edinburgh EH1 3QB
Scotland

ISBN 1 903765 49 8



British Library Cataloguing in Publication Data
A catalogue record for this book is available from the British Library

Typeset in 11/15pt Adobe Garamond with Franklin Gothic display
Design and production by Makar Publishing Production
Cover design by Mark Blackadder
Printed and bound in Great Britain by Cromwell Press

# Contents

# Preface

With this book, Hakuyu Okada has provided a great service, for there has never been a book-length history of clastic sedimentology. It is especially good to have the first coverage of Japanese sedimentology accessible to an international audience. The author also pro-vides more treatment of Russian contributions than generally appears in western literature. His international experiences make Okada perhaps uniquely qualified to author such an ambitious book.

This is not a textbook of sedimentology, rather it shows how the understanding of clastic sediments has evolved during the past three centuries within the larger context of stratigraphy and marine geology. The book is arranged in nine chapters. The author provides in the Prologue a summary of his aims for each chapter, which should be read before proceeding farther. With the exception of the Introduction, each chapter considers the chronological development of a particular aspect of the book's subjects. Chapters 2 and 3 provide a foundation in stratigraphy and basin analysis, including sequence stratigraphy, for all of the succeeding, more strictly sedimentological chapters. Topics covered by the later chapters include sedimentary petrology (covering mineralogy and texture), lithology (Walther's Law of facies and facies analysis), and the importance of ocean floor studies for sedimentology (from the *Challenger* Expedition to deep-sea drilling). Chapter 8 provides a concise history of stratigraphy and sedimentology in Japan, which enumerates the principal people involved in Japanese sedimentology over the years. The final chapter tries to anticipate future developments, including novel sections about the possibilities for the study of extra-terrestrial sedimentology and the environmental and social impacts of sedimentology.

After his book appeared in Japanese in 2000, sedimentologists from several nations urged Professor Okada to provide a more widely accessible English version, and we can be very grateful for his prompt response to that suggestion. Although Okada himself has a very good command of English, he solicited from several native-speaking colleagues critical reviews of the manuscript to assure clarity of the translation as well as of content and organisation. The result is a model for a very readable and interesting historical account of a field. Younger sediment-ologists will find Okada's book especially valuable for gaining a perspective of the development of clastic sedimentology, mucn of which their older colleagues participated in during the particularly important formative decades of the 1950s, 1960s, and 1970s. That older group will also find much of value, especially about contributions from regions of the world far from their own.

Okada's scholarship is very impressive. He has personally reviewed an enormous amount of literature from all over the world, has consulted colleagues across the globe, and has provided liberal referencing of the widely scattered literature on his subject. He has also reproduced many original illustrations and has provided portraits of many key, influential sedimentologists. The author's grasp of his subject and his integration of sedimentology with stratigraphy is truly outstanding and sets a challenging precedent for future scholars to strive for.

Hayuku Okada is an internationally famous sedimentologist with an exceptionally broad background, which prepared him admirably to undertake such an ambitious project as the global summary of the history of a complex field of study. He is presently Professor Emeritus of both Shizuoka and Kyushu Universities. In 1964–1965 he held a British Council Research Fellowship at Reading University in the United Kingdom. This experience provided him with an all-important introduction to the European sedimentological community and the opportunity to become fluent in English. Following that experience, Okada also travelled widely in North America where he made the acquaintance of many more sedimentologists. In addition he has visited China, Southeast Asia, India, Australia and New Zealand, the Philippines, Egypt and Sierra Leone. Okada has participated in several marine expeditions as well. All of these experiences have helped him to gain first-hand knowledge of global geology and to make him truly an international figure and an important scientific ambassador for Japan.

Hakuyu Okada received the Japan Prize of the Geological Society of Japan in 1993 and he served as President of that Society from 1994 to 1996. He has served as a Council Member of the International Association of Sedimentologists (1974–1982), has been Project Leader of an International Geological Correlation Project about 'Cretaceous Environmental Change in East and South Asia' (1993–1998), and is currently a Vice-President of the International Commission on the History of the Geological Sciences. He will serve as Honorary Chairman of the 17th International Sedimentological Congress scheduled to meet in Fukuoka, Japan in 2006.

Robert H. Dott, Jr.
*University of Wisconsin*
*December, 2004*

# Prologue

Sedimentology, still steadily developing as a basic discipline of geological sciences, was established as late as in the 50s and 60s of the twentieth century. However, there was a precursory stage as long as the history of geology itself before the establishment of modern sedimentology. The history of sedimentology, in a broad sense, gives us many interesting examples which help us to understand the development of any branch of science, because it attests, in many respects, to the paradigm theory of Thomas Kuhn (1962): namely that science evolves explosively (Miyashiro, 1998).

During the five decades since the establishment of sedimentology, many useful textbooks have been published either on its principles or on its methodologies. It is clear that they have established sedimentology as a systematized discipline. Those publications vary from introductory guides for beginners to specialized subjects for advanced students and researchers, but there has been no book that examines sedimentology from its genesis, through its subsequent development and to its present state, although interesting reviews of the history of sedimentology have recently been presented by R. H. Dott, Jr. (1987), E. Seibold and I. Seibold (2002) and G. V. Middleton (2003a).

This book sets out to fill this gap and the author will describe sedimentology on the basis of its historical evolution as a fundamental field of earth sciences. While it will focus on clastic sedimentology it is not intended to be a text-book of clastic sedimentology, many such books are readily available. Nevertheless the book will provide students, general readers and specialists with a clear appreciation of the contribution of sedimentology to the science of geology. The author first published the book in Japanese in 2002 (Okada, 2002), and in order to make the book accessible to a wider audience this English edition has been prepared, with the assistance of Alec Kenyon-Smith, in a revised and expanded form.

Since the book is concerned with ideas and their acceptance, it is necessary for the author to say something about how he made his selections. It is sometimes difficult to ascribe an original idea to one person and his co-authors. Most ideas in science are the product of an evolutionary process. This can be illustrated by one example: the recognition of the value of current and graded bedding (see Chapter 2). It is usual in textbooks to give credit to Edward Bailey – indeed the author does so in this book. But the story is not so simple. It would appear that Bailey did not at first recognize the value of crossbedding or graded bedding for determining younging direction – indeed he resisted the idea for several years. In 1924 two young graduates of the University of Wisconsin accompanied the Norwegian geologist Thorolf Vogt on a field excursion to the Highlands of Scotland and visited the complexly folded area at Ballachulish which had been mapped by Bailey. Bailey was not sure of the younging direction but the two students, who had been taught about younging criteria, immediately saw the correct younging direction and thus knew that Bailey's sequence was overturned (Dott, 2001). Vogt was convinced and wrote to Bailey, who remained sceptical. In 1927, Bailey accompanied a Princeton University field trip to the Archaean of Canada. The Canadian geologist T. L. Tanton demonstrated the value of both cross bedding and graded bedding in unravelling geologically complex areas. Then in 1929 the Princeton field trip was to the Highlands of Scotland and was led by Bailey. Tanton accompanied the group and at Ballachulish he was able to convince Bailey of the value of younging criteria and prove Bailey's succession was inverted. Bailey was convinced on the spot! He then invited Vogt to write a paper on the reinterpretation of the Ballachulish area and Tanton to write on the various criteria of younging. These, together with a paper by Bailey, appeared in the *Geological Magazine* of 1930. Later generations of geologists have perhaps awarded Bailey more credit than he deserved. (The author is indebted to Professor Robert H. Dott, Jr. for the foregoing account.)

In this case and in other cases in this book, the author has been influenced by the ideas and opinions he has absorbed throughout his geological career spanning half a century. He readily accepts that his readers may have other priorities but would say in his own defence that what follows reflects his own ideas and that he has taken care to present them as clearly and correctly as possible. He welcomes correspondence from interested persons.

The following are the author's aims for each chapter.

Chapter 1 gives the definition of sedimentology, and explains the objectives of sedimentological studies. It implies that this discipline is not only the most important field in geological sciences, but also reveals its strong influence upon such fields of the earth sciences as volcanology, petrology, mineralogy, economic geology, marine geology, and geomorphology, as well as upon such broad applied fields as the environmental sciences, geotechnology, civil engineering, ecology, soil sciences, and so on.

Chapter 2 is intended to make clear how sedimentology established its position during the 200-year history of geology when its nature was hidden within geology, and how its scientific form was germinated in the course of the development of stratigraphy.

Chapter 3 focuses on the development of observations of the depositional features of sediments, one of the most fundamental subjects of sedimentology. In particular, concepts of experimental sedimentology, megasedimentology, seismic stratigraphy, and sequence stratigraphy are described, consequent upon the establishment of these research methods.

Chapter 4 discusses the importance of sedimentary petrology as a major framework of sedimentology by reference to the textures and compositions of sediments, to grain morphometry and granulometry, all of which have contributed much to the exploitation of hydrocarbon resources.

Chapter 5 summarizes the concepts of facies and lithology, and discusses their development. It is stressed that the lithology concept was strongly developed in Russia, where a unique study method of sediments, not found in Western Europe and North America, was established. The reason for this is discussed in this chapter. The lithology concept became deeply rooted in Japan much earlier than in Western Europe and America.

Chapter 6 shows the final stage of the establishment of sedimentology as a new discipline of earth sciences. It begins with the evolutionary phase of the study of sediments and distinguishes the contributions of Western European, Russian and American schools. It was the amalgamation of these three schools that brought about the foundation of the new science 'sedimentology'. There were, however, serious arguments over the usage of the word 'sedimentology' before its general acceptance.

Chapter 7 describes how the study of ocean-floor sediments was indispensable to the steady development of sedimentology, referring to the great contributions made by both the *Challenger* Expedition in 1872–1876 and the Deep Sea Drilling Project (DSDP) started in 1968. The latter's achievements cover a range of subjects, but in this chapter the following three sedimentology-related topics have been selected as illustrations: the establishment of plate stratigraphy, the salinity crisis in the Mediterranean, and the black shale problem. It is, of course, universally accepted that the sedimentological contributions to the study of the ocean floor by the Deep Sea Drilling Project did much to establish the plate tectonics paradigm.

Chapter 8 introduces, without apology, the establishment and development of sedimentology in Japan, since it would appear that the contributions made by

Japanese geologists to sedimentary geology are not widely known in the English-speaking world. Sedimentology developed in Japan as part of stratigraphy, but its establishment was much earlier in Japan than in Europe. The detailed history of sedimentology in Japan is reviewed with critical analyses of recent developments.

Chapter 9 discusses new directions of sedimentology in the twenty-first century, which include the reunion of sedimentology and stratigraphy, extra-terrestrial sedimentology, and the urgent necessity of establishing environmental sedimentology and social sedimentology as significant branches of the subject.

In order to praise innovators in sedimentology, the portraits of as many as possible are included throughout this book in the belief that many readers will appreciate the opportunity to see some of the faces of those innovators.

# Acknowledgements

The Japanese edition of this book was published in 2002 and has been well received. Many people, not only in Japan but also from overseas, strongly urged the author to prepare an English edition, and in this respect, inspiration and encouragement were provided by Prof. P. Allen and Prof. J. R. L. Allen of the University of Reading, Prof. G. V. Middleton of McMaster University, Prof. A. J. Kenyon-Smith previously of Royal Holloway – the University of London, Dr. H. G. Reading of the University of Oxford, Prof. R. H. Dott, Jr. of the University of Wisconsin, Prof. J. McKenzie of ETH-Zentrum, Zurich, President of the International Association of Sedimentologists (IAS), Prof. R. Matsumoto of the University of Tokyo, Vice President of IAS and Chairperson for ISC 2006, Prof. J.-P. Calvo of Universidad Complutense (Madrid) and IAS (International Association of Sedimentologists) General Secretary, and Prof. D. R. Oldroyd, Vice President of the International Commission of the History of Geological Sciences (INHIGEO) as well as by Prof. T. Matsumoto, Member of the Science Academy of Japan, Prof. M. Omori of Azabu University, Prof. Y. Takayanagi of Tohoku University, Prof. A. Miyashiro of New York State University, Prof. S. Uyeda, Member of the Science Academy of Japan, Prof. T. Shiki of Kyoto University, and Prof. S. Mizutani of Nagoya University. The translation was also critically reviewed and improved by Prof. Alec J. Kenyon-Smith of the University of London and by Prof. R. H. Dott, Jr. The author also acknowledges Prof. R. H. Dott's readiness to provide a Preface to this work. Thanks are also given to Mr Anthony Kinahan of Dunedin Academic Press in Edinburgh and Mr Nobuo Sekita of Kokon Shoin, publishers, Tokyo for their support and encouragement. Finally, the author wishes to thank his wife Yoko Okada for her constant support.

My sincere thanks are also due to Prof. E. Mutti of the University of Parma, Italy and Prof. G. B. Vai of the University of Bologna, Italy, for a portrait of Nicolaus Steno, Prof. Algimantas Griglis of the Vilnius University in Lithuania for a portrait of Abraham Gottlob Werner, Prof. Gerhard Einsele of the University of Tübingen, Germany, for a portrait of Johannes Walther, Prof. J. Aubouin, Member of the Science Academy of France, for a portrait of Lucien Cayeux, and Prof. G. V. Kirillova of the Institute of Tectonics and Geophysics, Far East Branch of Russian Academy of Sciences, for a portrait of N. M. Strakhov.

Hakuyu Okada
*Fukuoka, Japan*
*November 2004*

# List of figures

# 1

# Introduction

## 1.1 What is sedimentology?

Sedimentology is a discipline of the earth sciences that studies the nature of sediments and sedimentary rocks and the processes of their formation in order to determine the earth's past environments. Hakon Wadell, who proposed the word 'sedimentology' in 1932, gave a wide and flexible definition to sedimentology, namely that 'sedimentology is a scientific study of sediments.' (Wadell, 1932b.)

The formational processes of sediments are controlled collectively by physical, chemical and biological events in the earth's surface environments under the influence of solar energy and earth gravity, being mutually influenced by the lithosphere, hydrosphere, atmosphere and biosphere. In other words, sedimentology attempts to reconstruct earth environments on the basis of conceptualized depositional conditions as influenced by changing processes through weathering, transportation, deposition and diagenesis. Sedimentology, therefore, plays important roles not only in the search for and development of such natural resources as petroleum, natural gas and coal, but also in prospecting for sedimentary metallic ores. In addition, sedimentology provides fundamental information and methods for the prediction and mitigation of natural disasters as well as for assessments of changed earth environments caused by artificial developments.

The research methods of sedimentology are summarized in Figure 1.1.

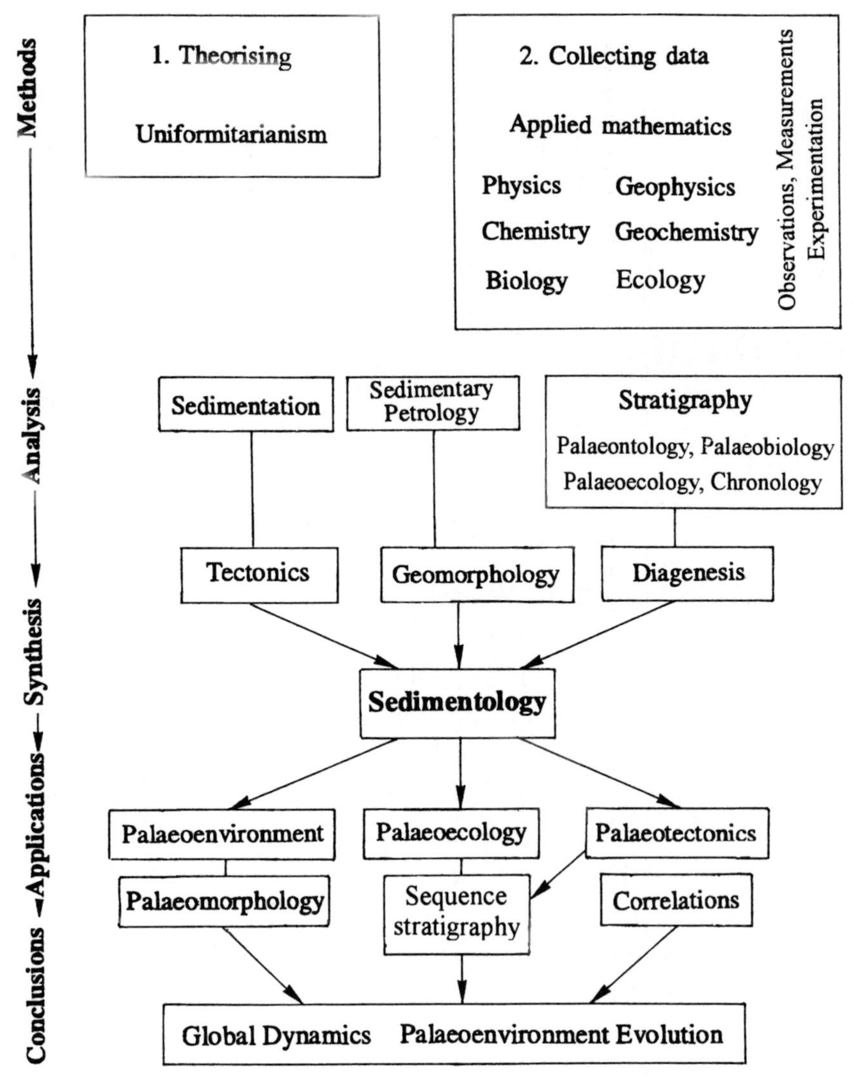


Fig. 1.1 Diagram showing the methodology of sedimentology. The framework of the diagram is based on Lombard (1978).

## 1.2 Sedimentology and its related sciences

Sedimentology is a fundamental field of the geological sciences constructed on the study of sediments and sedimentary rocks based on stratigraphy, historical geology and structural geology. (The word 'geology' comes from ancient Greece, and the earliest published reference to geology was by Erasmus Warren in 1690 (Woodward, 1911).) Furthermore, sedimentology is also significant for the study

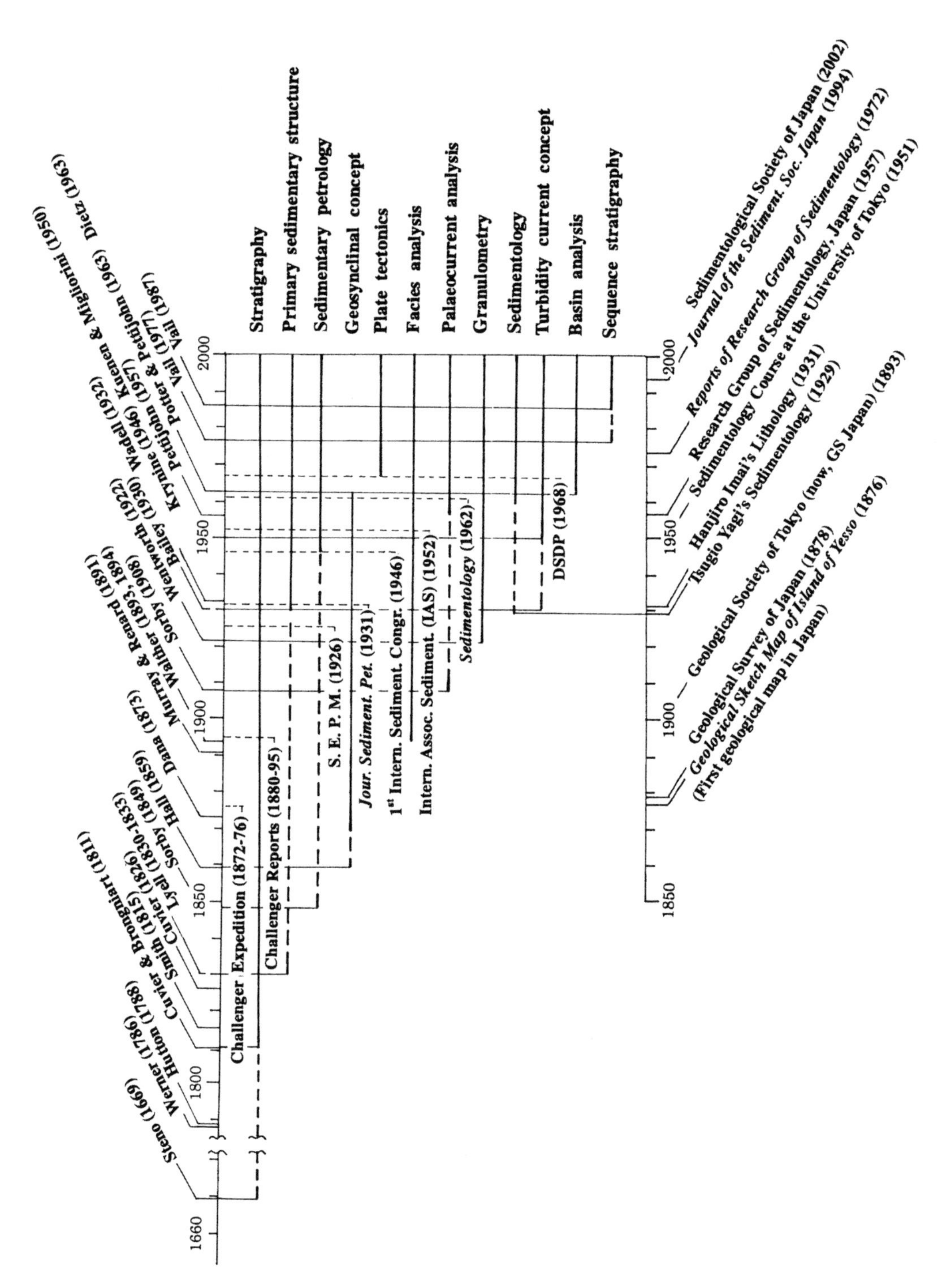


**Fig. 1.2** Summary of the development of sedimentology. The bottom part indicates the history of Japanese sedimentology.

of fluid-related phenomena of oceans, lakes, snow and glaciers as well as in the ejection, transportation and deposition of volcanic ejecta, genesis of soils, landslides and many other examples. Sediments and sedimentary rocks record data of environmental changes of the earth, including changes brought about by human activities. As thus defined, sedimentology is, in a broad sense, an environmental science. It is, however, suggested that a discrimination is necessary between environmental sedimentology, which treats the analysis and reconstruction of natural environments, and *social sedimentology*, which is concerned with artificial environments and, when necessary, appropriate countermeasures.

Sedimentology, consequently, is not restricted within geological sciences as a wide discipline such as stratigraphy, petrology, mineralogy, palaeontology, and so on, but covers much larger fields including geomorphology, volcanology, oceanography, glaciology, limnology, soil science, coastal engineering, and others. As a recent trend, geology-based sedimentology is called sedimentary geology, in which emphasis lies on the interpretation of the environments of sedimentary basins by the study of the origins of sediments and the three-dimensional analysis of the distributions of strata and sedimentary structures. In this sense it is clear that sedimentology is closely related to stratigraphy.

Though sedimentology is a basic field of earth sciences, the history of its establishment is quite young. Only in the early 1950s was sedimentology established as a discipline of the earth sciences in its own right. As the details of its development are described in the following chapters, it is quite interesting to note that the period of the establishment of sedimentology in the 1960s corresponded to that of the great revolution of paradigms from the geosyncline to plate tectonics (Miyashiro, 1998) (Fig. 1.2).

# 2

# Pre-sedimentology: development of geology and stratigraphy

The study of strata before the establishment of sedimentology was itself the history of geology. Therefore, we must be discuss here the significance and purpose of the study of strata which supported the development of modern geology.

## 2.1. Accumulation of strata: the recognition of geological time

The first step in the birth of modern geology was the realisation of the role of time in the accumulation of strata. The study of strata helped in understanding the processes of the sequential evolution of the earth. The first contribution to this study was the recognition of stratification by Nicolaus Steno (1638–1686) (Fig. 2.1) (Steno, 1669), and it is said that Steno described the three basic principles of the deposition of strata: namely that strata are deposited horizontally, are laterally continuous, and lie successively on earlier strata. These were to be called the three Laws of Stratigraphy (Fritz and Moore, 1988). They are now expressed as the Law of Original Horizontality or the First Law of Deposition of Strata, the Law of Lateral Continuity or the Second Law, and the Law of Superposition or the Third Law. These three laws offered the fundamental basis for the development of stratigraphy. In addition to this, and not less important, is that Steno recognized the time concept in the succession of strata. This is fundamentally important in geology (Cipriani, 1992; Yamada, 2003).

**Fig. 2.1.** Nicolaus Steno (*reproduced by permission of Prof. G. B. Vai, University of Bologna, Italy*).

In spite of his correct recognition of strata, Steno could not contribute more to geology as he failed to escape the established religious doctrine. Furthermore, the development of geology had to remain stagnant, chiefly because the distinction between sedimentary rocks and igneous rocks was not recognized at that time. To change such a situation, Abraham Gottlob Werner (1750–1817) (Fig. 2.2) made a great contribution in considering that all the rocks at the earth's surface were hydrogenic (water-related), suggesting that even granite was precipitated

**Fig. 2.2.** Abraham Gottlob Werner (*reproduced by permission of Prof. Algimantas Griglis, Vilnius University, Lithuania*).

from the sea. He defined his scientific method as '*geognosy*' and thus made Steno's recognition of strata fundamental. According to Werner, geognosy is defined as the science dealing with the origin and mutual relationship of the rocks and minerals that constitute the earth.

Werner (1786) classified all rocks on the earth into the following four series (Fritz and Moore, 1988; Shimizu, 1996):

Alluvium
Volcanic rocks
Flöts
Primary series (Primitive series)

The Primary series, the oldest, consists of granite, gneiss, slate, marble and schist, and contains no fossils. Werner considered that all of them were deposited chemically from the primitive ocean. According to him, the overlying Flöts contained fossil-rich limestone, conglomerate, sandstone, coal, chalk and gypsum, all of which were also chemical deposits. Volcanic rocks were interpreted as products of melting by burning coal in the Flöts rocks. The uppermost, Alluvium, was regarded as the deposit produced by erosion and deposition of

**Fig. 2.3.** Robert Jameson (Linklater, 1972).

the underlying rocks. It is important here to note that Werner demonstrated a stratigraphic succession on a global scale, illustrating the concept of a standard stratigraphy.

Since the hydrogenic concept of Werner was not against the Biblical doctrines as represented by the Noah Deluge, it became widely accepted and those people supporting it were called Wernerians or Neptunists. The view that almost all rocks had a water related origin was supported by Robert Jameson (1774– 1854) (Fig. 2.3), a student of Werner's and professor of Edinburgh University who had a major influence in Edinburgh society (Jameson, 1808). He could not, however, have made his philosophy clear without the work of James Hutton, a genius of the Edinburgh enlightenment.

## 2.2. Birth of modern geology: the establishment of the unconformity concept

The concept of geognosy, which was proposed by Werner for all the rocks of the earth, was strongly connected with an aquatic environment, whereas evidence had made it clear that granites as well as basalts had a magmatic, molten rock, origin. At the end of the eighteenth century, there was a leading figure in Edinburgh (then a centre of the Wernerian school, the chief centre being Freiburg, Germany) who visited many parts of Scotland and carefully observed rock exposures. He was James Hutton (1726–1797) (Fig. 2.4). He described his detailed observations of intrusions

**Fig. 2.4** James Hutton, detail, by Sir Henry Raeburn (*reproduced by permission of the Scottish National Portrait Gallery*).

of granite in western Scotland, which thermally altered rocks around the intrusion. He also described the relationship between the basalt of Arthur's Seat, a prominent hill in Edinburgh, and the surrounding Old Red Sandstone showing that the Seat basalt, rather than being deposited from water, is younger than the underlying and overlying sedimentary rocks, because, as it intruded, it had heated and altered blocks of the sedimentary rocks above and below it. Hutton had observed igneous rocks correctly and revealed the time-related relationship of geological phenomena as the basis of field geology (McIntyre and McKirdy, 1997).

In the period of Werner and Hutton, those who supported the observations of igneous rocks by James Hutton, were called Huttonians, Plutonists, or Vulcanists.

James Hutton also observed an unconformity, an abrupt angular discordance between strata, in the outcrops at Jedburgh and at Siccar Point, in the eastern Southern Uplands of Scotland. Through these correct observations of geological phenomena, he constructed a basis for modern geology (Hutton, 1788). As Siccar Point is located by the shoreline, the unconformity between the underlying vertically inclined Silurian turbidites, deep-sea sediments, which Hutton called the *schistus*, and the overlying flat cover of the Devonian Old Red Sandstone non-marine sediments is clearly visible. Hutton vividly described his significant observation of the Siccar Point unconformity in his book *The Theory of the Earth* (1795) and he realized that the unconformity represented a buried, ancient landscape. This outcrop is today well-known as a World Heritage Site (Fig. 2.5). The term unconformity is now classified into angular unconformity, nonconformity, disconformity and paraconformity, in which the Huttonian unconformity is classed as an angular unconformity.

Hutton proposed the origin of an unconformity as follows: that strata originally deposited horizontally on the sea floor became lithified, deformed, and uplifted, then eroded and subsided, and later covered by younger sediments thus making the unconformity. In this concept, there is a combination of geological phenomena such as deposition of strata, folding, uplifting, erosion, subsidence, and then deposition of more sediments. He also noted that, including periods for which there is no evidence, an enormously long time is necessary to produce such an unconformity cycle. Furthermore, each process for producing the 'unconformity' is also operating at present. Thus, the concept of uniformitarianism was born (*see* Chapter 2.4).

**Fig. 2.5** Hutton's unconformity exposed at Siccar Point about 50 km east of Edinburgh (photo by H. Okada). The vertical underlying strata are the Silurian flysch turbidites, and the flat cover is the Devonian Old Red Sandstone.

## 2.3 Establishment of stratigraphy: the geological map and the correlation of strata

An idea of a standard stratigraphy created by Werner in 1786 had by the early nineteenth century evolved into modern stratigraphy through the recognition of the value of fossils. This discipline was developed almost contemporaneously by a vertebrate palaeontologist, Georges Cuvier (1769–1832) (Fig. 2.6), a professor of the Collège de France, a palaeobotanist, Alexandre Brongniart (1770–1847) in France, and an English engineer William Smith (1769–1839) in Britain.

Georges Cuvier, who was studying the stratigraphy and fossils of Palaeogene strata occurring around Paris, described for the first time the details of the stratigraphy of the Paris Basin in collaboration with Alexandre Brongniart, who reported on the plant fossils in this sequence, in 1808. They described the characteristic lithologic features of each part of the succession and the changes in contained fossils. Because strata containing non-marine animal fossils and those with marine animal fossils were repeatedly intercalated with each other, Cuvier and Brongniart recognized that abrupt changes from marine to non-marine environments caused differences in the contained fossils. They suggested

**Fig. 2.6** Georges Cuvier (*reproduced by permission of Dr. Anne-Marie Bodergat, Université Claude Bernard, France*).

that the differences in both the nature of the sediments and assemblages of fossils took place due to repeated sudden changes of earth environments, which they called *révolutions*. Though this paper included no illustrations, Cuvier published a revised and expanded edition of it in 1811 with the title of *Essai sur la geographie mineralogique des environs de Paris*, in which eleven differentiated strata in the Paris Basin, and their stratigraphy, were illustrated with a lithologic column (Cuvier and Brongniart, 1811). Fig. 2.7 shows the lithologic column of the Eocene sequence in the Paris Basin (Cuvier and Brongniart, 1822). This figure indicates

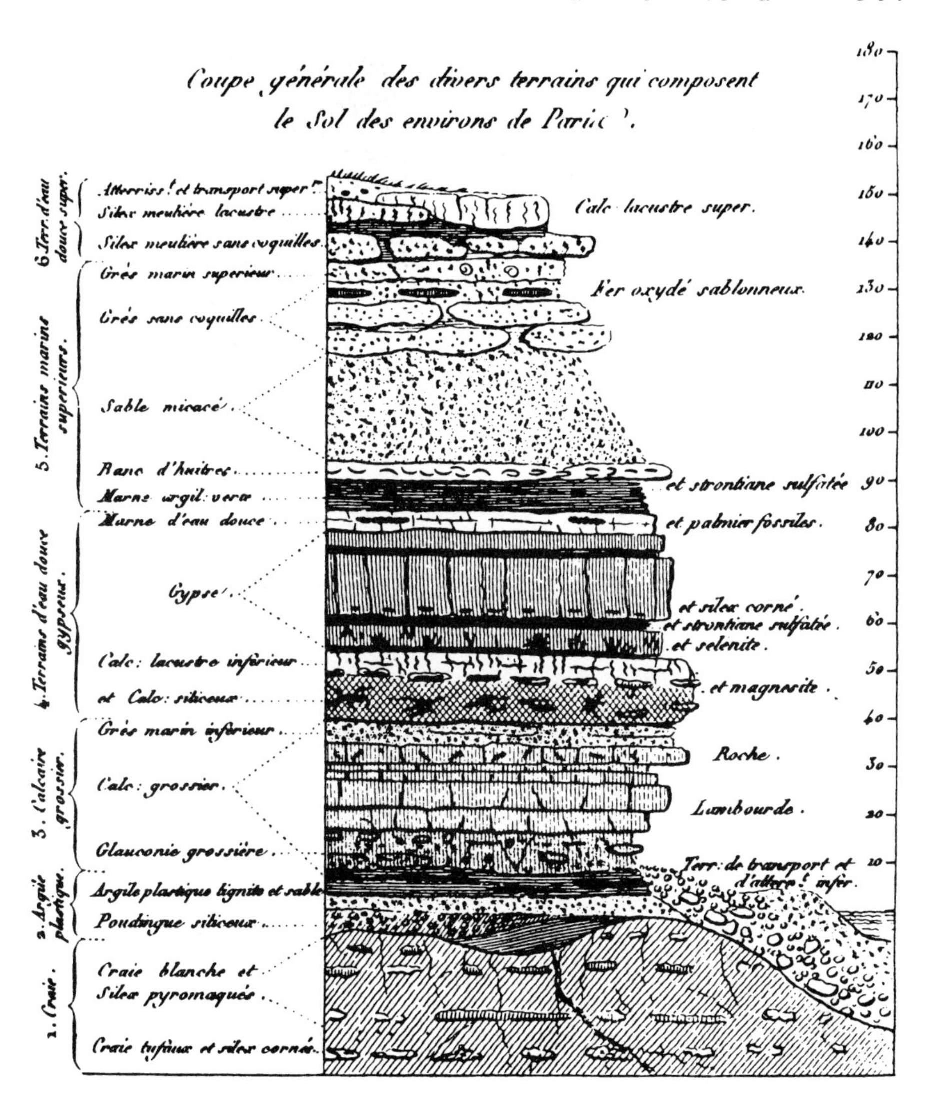

Fig. 2.7 Lithologic column of the Eocene of the Paris Basin presented by Georges Cuvier and A. Brongniart (Ager, 1993). 1: white chalk with flints, 2: sandy, lignitic clay with siliceous conglomerate, 3: lower marine sandstone, 4: gypsiferous freshwater beds with marl and gypsum, 5: upper marine beds with sandstone, marly clay, and oyster beds, 6: upper freshwater beds with lacustrine limestone. (*Reproduced by permission of Cambridge University Press.*)

quite impressively and vividly the observed facts, suggesting that stratigraphic records may point clearly to rapid changes of earth's environments.

It should be emphasized that the book also contains a coloured 1/200,000 geological map of the Paris Basin accompanied by geological profiles in different scales. Based on these achievements, Cuvier later published a book entitled

**Fig. 2.8** William Smith (Woodward, 1911).

*Discours sur les revolutions de la surface du globe*, stating that repeated revolutions of the earth's surface caused the appearance of different fossils in each part of the stratigraphic sequence (Cuvier, 1826). This concept was later called *catastrophism* (Whewell, 1832) and he is thus regarded as its proposer.

Working almost at the same time as Cuvier and Brongniart, William Smith (1769–1839 (Fig. 2.8) published in 1815 a coloured geological map based on his own surveys, in which he illustrated not only the whole of England and Wales but also the southern part of Scotland and a part of Ireland on the scale of 5 miles/1 inch (approximately 1/320,000) (Fig. 2.9 and *see* frontispiece). In this map 24

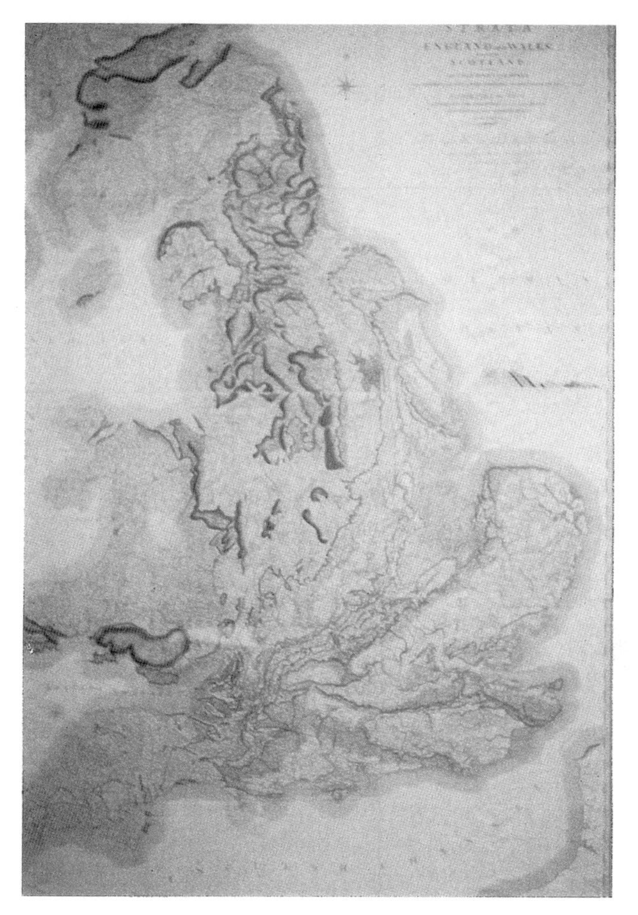

**Fig. 2.9** Monochrome copy of the geological map of England, Wales and part of Scotland by William Smith, 1815 (*see also* frontispiece) (© *The Natural History Museum, London*).

stratigraphic divisions were indicated by different colours, each of which showed accurately their areal distribution and the stratigraphic relationship between them. At the same time, a geological section showing the successive relationships from the London Clay (in the southeast of England) at the top to the granite and gneiss in western Wales at the bottom was attached (Fig. 2.10).

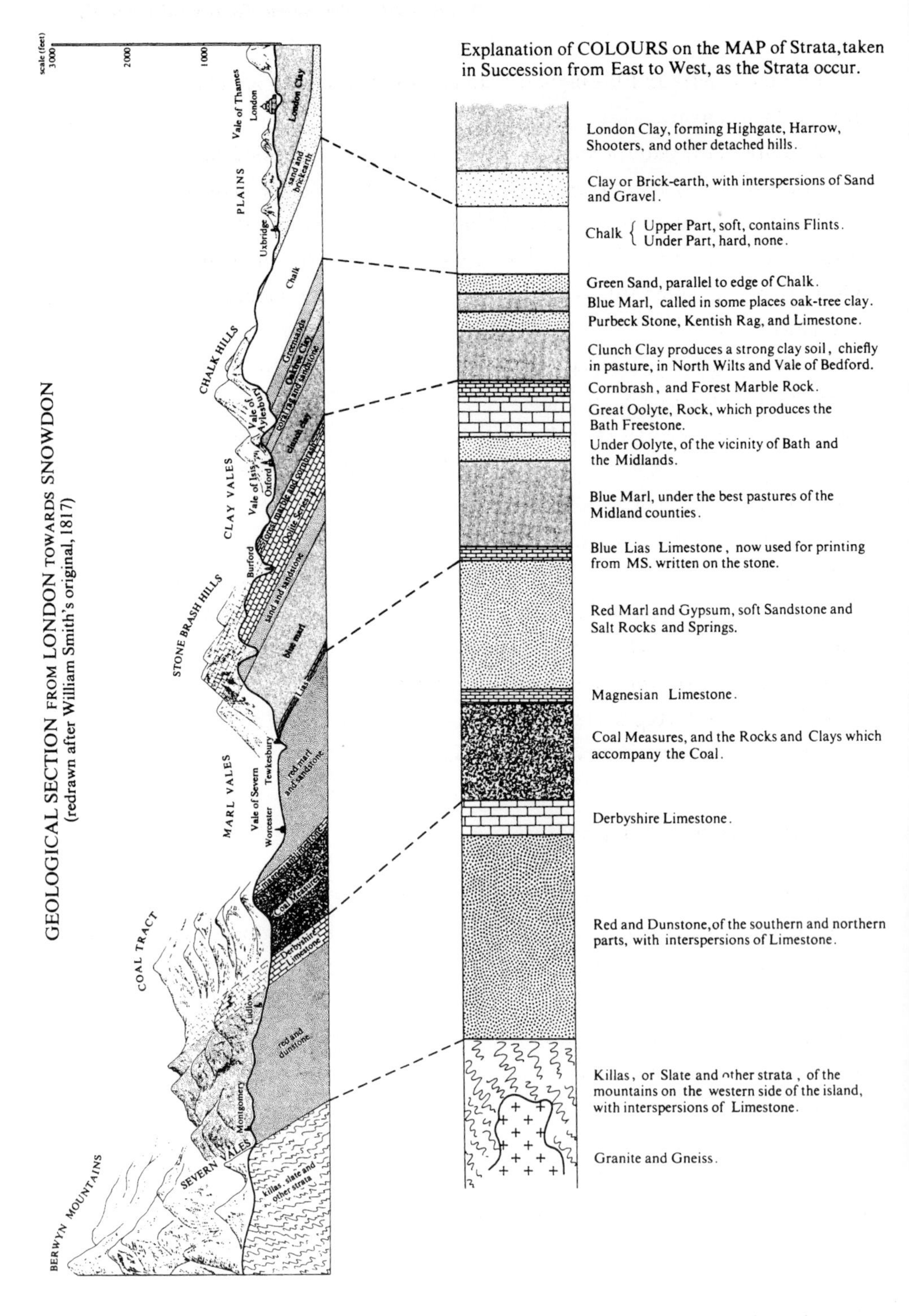


**Fig. 2.10** Geological section of England and Wales by William Smith and a columnar succession of strata redrawn by Wilson (1994, fig. 4.8). (*Reproduced by permission of The Geological Society.*)

Both the Cuvier and Brongniart map and the Smith map are two of the very first geological maps.

In addition, the publication by William Smith in 1816 entitled *Strata identified by organized fossils, containing prints on coloured paper of the most characteristic specimens in each stratum* was quite unique, in that particular strata, containing distinctive kinds of fossils, were illustrated by the help of coloured plates. This illustration was intended as a practical application to identify strata using contained fossils.

Producing geological maps and identifying strata with fossils were the outstanding achievements of Smith, a civil engineer. They were significant not only for the development of geology but also for engineering works. In particular, his geological maps were very important for prospecting and mining coals, the principal force of the First Industrial Revolution.

Thus, Cuvier, Brongniart and Smith established the Law of Identification of Strata by Fossils which states that strata in different areas and with different lithologies can be identified as equivalent if contained fossils can be correlated (Brongniart, 1821). Such a stratigraphic concept was strengthened by introducing the concept of fossils zones and by establishing fossils as the indicators of relative times by A. D. d'Orbigny (1842) and Albert Oppel (1856–1858). This, in due time, led to the present discipline of biostratigraphy.

## 2.4 Uniformitarianism

At the time of James Hutton the age of the Earth was accepted as quite young. James Ussher, an archbishop of the Church of England, pondering the creation of the Heavens and the Earth in his book *Annals of the World* (1650–1654), stated that the Earth was born from dusts on the Friday night of 23 October 4004 B.C. His conclusion was to become the accepted ecclesiastic view. Society at that time was so controlled by church doctrine that every phenomenon on the Earth was seen as a product of God's hand. Hutton, however, considered that the Earth was created far back in the past and that changes on the earth's face occurred in a natural order and over a long time with these processes still going on even at present. Nature on Earth has thus acquired its order and eternity, as is well expressed as 'In nature there is wisdom, system, and consistency' (Hutton, 1788, p. 304).

When Hutton's thinking was restated plainly by his friend John Playfair in the book *Illustrations of the Huttonian Theory of the Earth* (1802), it intensely influenced the world of thought in Edinburgh. The basic theory of James Hutton,

**Fig. 2.11** Charles Lyell (*reproduced by permission of The National Portrait Gallery, London*).

that earth's history is created by a procession of slow and small changes resulting in large and significant ones, was elaborated by Sir Charles Lyell (1797–1875) (Fig. 2.11). Lyell refined and developed the Huttonian Theory as a fundamental concept for, and methodology in, geology. This scientific concept is understood by accepting that processes and events proceeding at present on the earth's crust are applicable to the interpretation of past geological phenomena. This thinking put life to the golden saying that 'The present is the key to the past.' This thought is explicitly described in Lyell's masterpiece *Principles of Geology* published in 1830–1833. Indeed, the principles of geology are clearly presented on the front page of his book: 'An attempt to explain the former changes of the earth's surface, by reference to causes now in operation.'

Such a concept was named *uniformitarianism* by William Whewell (1832) of Cambridge University. At the same time, he introduced the word *catastrophism* as the counterpoint to uniformitarianism.

Lyell's uniformitarianism is still alive as a basic principle of geology (Lombard, 1978) (Fig. 1.1). It can be said that the interpretation of past sedimentary environments is carried out with reference to present earth environments. This methodology is called *actualism*. In this respect, Johannes Walther (1860–1937), a German biologist and geologist, coined the word *ontology* in order to emphasize that understanding present sedimentary facies is truly important for reconstructing past sedimentary environments (see Chapter 5.2).

Charles Lyell discussed the processes of formation of sediments in his book *Principles of Geology* in reference to detailed observations of various changes on the earth's surface. For this reason, M. R. Leeder (1998) proposed that Sir Charles Lyell should be recognized as a pioneer of sedimentology. However, as Akiho Miyashiro (1998) wrote, 'uniformitarianism has recently been criticized by suggestions that catastrophic geological processes and events were, and are, more important in earth's history than slow, steady ones'. Derek Ager proposed a concept of 'Catastrophic Uniformitarianism' in 1973 or 'New Catastrophism' in 1993, which, he insisted, explains changes of earth environments as recorded in the stratigraphic record more correctly.

## 2.5 The geosynclinal concept and the study of sediments

With the acceptance of the plate tectonics theory the history of geology can be divided into two major stages. The first is represented by the concept of the geosyncline, and the second by the plate tectonics concept. The development of geology due to the geosynclinal concept was achieved by the study of sedimentary rocks.

The idea of the geosyncline emerged with the study of the earth's large mountain ranges. It was well known in the late eighteenth century that the European Alps are characterized by large scale folded structures. It was, however, in the Appalachian Mountains of North America that the geological features of large mountains were first analyzed. James Hall (1811–1898) (Fig. 2.12), a geologist of the New York State Geological Survey at that time, determined, in

**Fig. 2.12** James Hall (Eckel, 1982).
*(Reproduced by permission of the Geological Society of America.)*

his survey of Palaeozoic strata exposed mainly in the state of New York, that the folded strata forming the Appalachians were extremely thick, whereas the time-equivalent strata exposed around the upstream area of the Mississippi River west of the Appalachians were quite thin, widely distributed and almost horizontal (Hall, 1859). Furthermore, a surprising feature was that, notwithstanding the contrasting difference in thickness, the lithologies were similar in that the strata in both areas were formed of quartzose sandstone, calcareous shale, and organic limestone. As a consequence, he considered that the folded strata forming the mountains and the horizontal strata on the western side were both originally deposited in shallow-sea areas, and later a basin subsided as the deposition proceeded on the mountain site. It followed that strata in the mountain area attained a considerable thickness and were later intensely deformed. His view that such a process must have been common in all other mountains on the earth made a strong impact on geological thought.

**Fig. 2.13** James D. Dana (Eckel, 1982). (*Reproduced by permission of the Geological Society of America.*)

Some years later, James Dwight Dana (1813–1895) (Fig. 2.13), a professor of Yale University, questioning how such thick accumulations of sediments as described by James Hall were developed, took account of the then widely accepted theory of a shrinking earth. According to this theory, the shrinking of the earth was caused by the cooling of the hot primitive earth, which consequently produced wrinkles in the earth's crust, thus creating many narrow elongated depressions in which thick sequences of sediments were accumulated. Dana (1873) called this depression mechanism 'geosynclinal'. This was the first use of the word 'geosyncline'. Thus the geosynclinal concept was established by Hall and Dana. This concept was immediately brought to Europe and applied to the study of the Alps. The study of geosynclines in Europe, however, was to develop in a different direction from that in North America.

The studies of the European Alps, for example by Eduard Suess (1875) and Melchior Neumayer (1875), both professors of Vienna University, showed for the first time that the geosynclinal sediments of the Alps were oceanic. In particular, Neumayer considered that Mesozoic radiolarian chert is comparable in origin to the radiolarian ooze of the modern Pacific and Indian Ocean floors. A background to this was, of course, the remarkable discoveries made aboard HMS *Challenger* in its worldwide ocean research expedition from 1872 to 1876 (*see* Chapter 7). On this voyage Sir John Murray made clear the distribution and nature of ocean-floor sediments (Murray and Renard, 1891), which included the discovery of radiolarian ooze. In the course of the study of the European Alps, the deep-sea origin of geosynclinal sediments gained a powerful position. As a consequence, a controversy developed between the shallow-sea nature of the geosyncline developed in America and the deep-sea origin advocated in Europe (Okada, 1971b).

Subsequently, Marcel Bertrand (1897) of the French Geological Survey proposed a concept of geosynclinal cycles through his study of the French Alps. It defines that the main rocks constituting geosynclines consist of the following four facies in ascending order: (1) *gneiss*, (2) *flysch schisteux*, (3) *flysch grossier*, (4) *grés rouges* (= molasse). The *gneiss* facies is the basement of a geosyncline; the *flysch schisteux* facies represents the central part of a geosyncline in its early stage; the *flysch grossier* facies represents sediments filling a geosyncline; and the *grés rouges* facies are the sediments deposited at the foot of newly uplifted mountains. Later Argand (1916) and Arbentz (1919) linked the concept of geosynclinal cycles to tectonic (orogenic) movements. It was van Watershoot van der Gracht (1931), who introduced the terms of *flysch* and *molasse* and the concept of orogenic development of geosynclines to North America.

The terms flysch and molasse were originally formation names in the Swiss Alps, however the terms were later widely applied to groups of rocks with similar characteristics in other orogenic belts of the world. The Flysch in the Swiss Alps consists of very thick sedimentary successions composed mainly of alternations of thinly bedded sandstone and shale ranging in age from Late Cretaceous to Oligocene. Such thick accumulations of alternating sandstones and shale are generally deposited in the geosynclinal stage of orogenic movements. It is a characteristic feature that individual beds of these sandstones show graded bedding. The Molasse in the Swiss Alps referred to fresh-water and shallow-sea sediments, deposited at the foot of mountains uplifted in late Oligocene to Miocene times, which consist of conglomerate, sandstone and shale with a variety of lithofacies. Thus, the notion became established that the flysch facies indicates the geosynclinal stage of the early orogenic cycle, whereas the molasse facies represents the post-orogenic stage.

The study of geosynclinal sediments was significantly developed through sedimentary petrology as modernized by Paul D. Krynine (1942) which is described in detail later in Chapter 4.2. At about the same time, the study of flysch deposits made a remarkable advance following the introduction of the concept of turbidity currents by Ph. H. Kuenen and C. I. Migliorini (1950). This is also discussed later (Chapter 3.4).

For a moment let us return to the general character of geosynclines. Émile Haug, a professor of the Sorbonne University in Paris, pointed out in 1900 that geosynclines in North America are usually located on the continental crust, whereas those in the European Alps are situated between two continents and are much larger in scale. In his study of black shale containing ammonites in the Jura Mountains in the border region between France and Switzerland, Haug noted that the lithologic features of geosynclinal sediments are characterized not only by thick monotonous shale, with no evidence of unconformities, and deep-sea sediments with pelagic fossils, but also by shallow marine deposits with distinct facies changes and shallow-sea fossils.

Haug's discoveries were refined in Britain, first by G. W. Tyrrell (1933), who indicated that geosynclinal sediments are characterized by greywacke sandstones, and by E. B. Bailey (1936) who emphasized the sedimentary structure of graded bedding in greywacke sandstones. These discoveries contributed much to the progress in the study of geosynclinal sediments. The development in Great Britain reached its acme through the work of O. T. Jones. In his presidential address on the evolution of a geosyncline given to the Geological Society of London in 1938, Jones summarized the characteristics and evolutionary history of the Welsh geosyncline (Jones, 1938). He recognized two contrasting facies in Wales and the Welsh Borderland: the Shelly Facies on the adjacent stable platform to the east and the Graptolite Facies in the geosyncline (Fig. 2.14). The Shelly Facies is characterized by clean quartzose sandstone, fossil-bearing limestone and fossiliferous shale, indicative of shallow seas, while in contrast, the Graptolite Facies shows very thick, monotonous successions of alternating beds of graded sandstone and dark shale. Fossils are generally rare, but graptolites are sometimes found in the shale, suggesting deep-sea environments. It should be noted that graptolites, being 'free floating' do not, in themselves, suggest a deep water environment. Indeed, because they are free floating they are also found in the shallow water successions of the Shelly Facies. Because they are so widely distributed and evolved rapidly they are excellent for time correlation. Jones made great use of this fact. He used the absence of shallow water fossils in the Graptolite Facies to indicate a 'deep sea' environment. Graptolites were almost the only fossils found, hence the name Graptolite Facies. Later it was recognized that the alternating sandstone and shale sequence indicates the typical flysch facies, and is characterized by turbidites (Wood and Smith, 1959). Also very significant was that Jones recognized contemporaneous (Ordovician) volcanic rocks in Wales associated with the geosynclinal sediments. This was a prelude for Stille's (and Kay's) eugeosyncline (Dott, 1974).

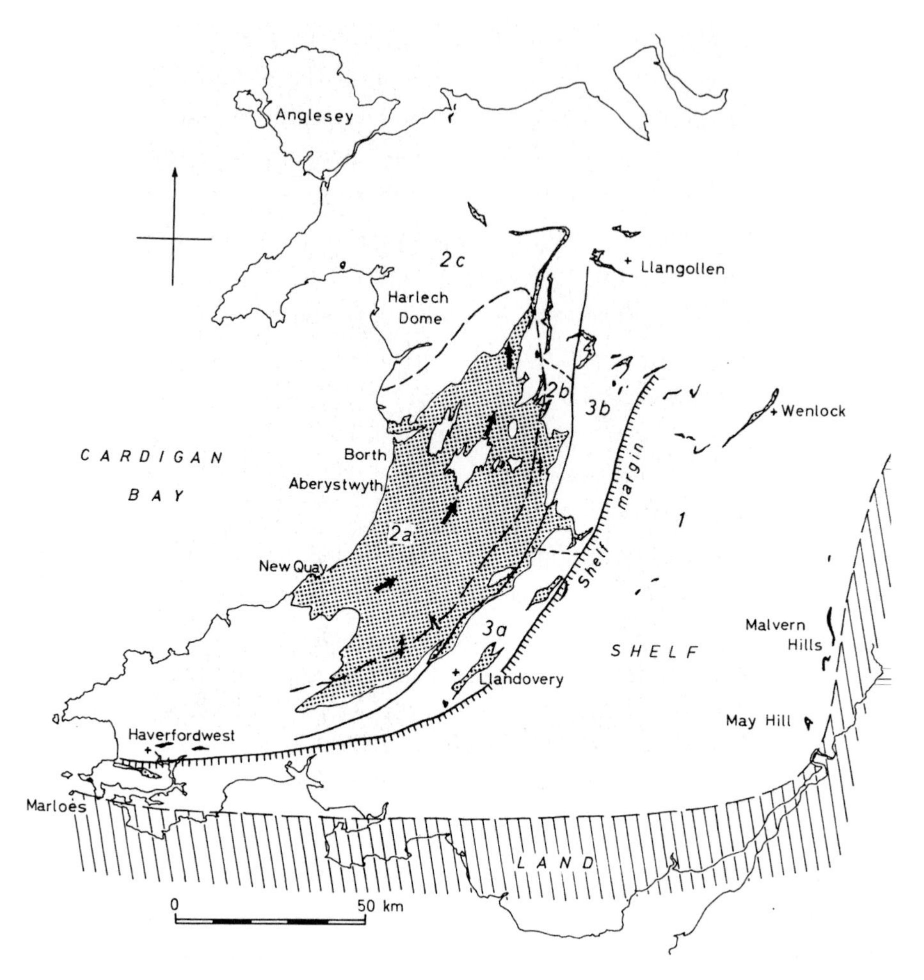


**Fig. 2.14** Lithofacies of the upper Llandoverian of the Lower Silurian in Wales (after Ziegler, 1965; Okada, 1967; and A. J. Smith's unpublished data). 1: shelf facies (Shelly Facies), 2: turbidite facies (Graptolite Facies), 2a: sandstone facies, 2b: siltstone facies, 2c: claystone facies, 3: marginal shelf facies, 3a: shelly facies, 3b: graptolite facies. Arrows: palaeocurrent directions. Dotted areas: distribution of the upper Llandoverian, Obliquely lined areas: land and basement area

Geosynclines in the orogenic belts of the world were studied systematically in the early twentieth century. From these studies the relationship between the stages of geosynclines and magmatism became apparent as was the significance of the timing of tectonic movements thus leading to the recognition of a close connection between geosynclines and orogenesis. For example, Leopold Kober (1933) differentiated the orogenic belts (*Orogen*) from stable regions (*Kraton*) of the earth's surface, and considered that a cross-section of mountains shows a

**Fig. 2.15** Marshall Kay (photo by H. Okada).

symmetric structure. Following Kober, Hans Stille separated the craton (*Kraton*) of stable areas from the geosyncline of mobile regions, presenting a system of geosynclinal orogenesis (Stille, 1936). He classified geosynclines into orthogeosyncline (*Orthogeosynklinalen*) accompanied by intensely deformed strata, magmatism and metamorphism, and parageosyncline (*Parageosynklinalen*), a filled depression having relatively gentle tectonic structures. Orthogeosynclines were further subdivided into eugeosyncline (*Eugeosynklinalen*) with igneous activity and miogeosyncline (*Miogeosynklinalen*) without magmatism. Stille's classification of geosynclines gave a strong impetus to the study of geosynclines.

Marshall Kay (1904–1995) (Fig. 2.15), a professor of Columbia University, extended the geosynclinal orogenesis concepts of Kober and Stille with a new classification system based on his intensive study of geosynclines in North America (Kay, 1951). According to him, geosynclines basically consist of two parts – a couplet: the eugeosyncline and the miogeosyncline (Fig. 2.16). The eugeosyncline is characterized by extremely thick clastic rock sequences and igneous activity, while the miogeosyncline, adjacent to the continent, is

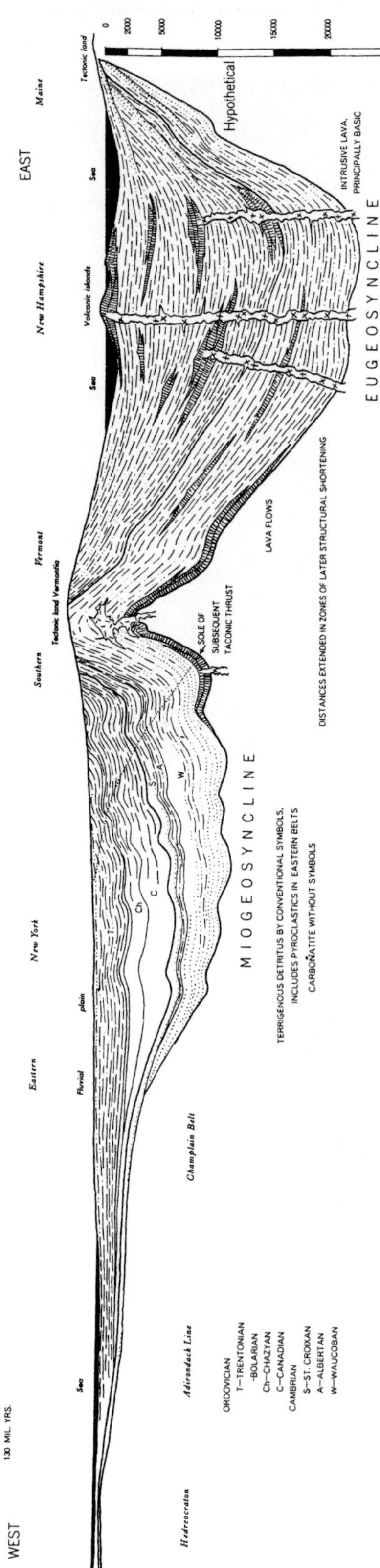


**Fig. 2.16** Kay's model of geosynclines (Kay, 1951) showing the typical eugeosyncline and miogeosyncline couplet in the Cambrian–middle Ordovician section from New York to Maine. (*Reproduced by permission of the Geological Society of America.*)

characterized by carbonate sediments without igneous activity. As a source for clastic sediments in the eugeosyncline, he proposed hypothetical tectonic uplands and volcanic islands within the eugeosynclines.

It was Jean Aubouin (1965) (Fig. 2.17), a professor of Pierre et Marie Curie University in France, who proposed the final scheme of geosynclines; the last in the history of their studies. His scheme was basically similar to the Kay model of geosynclines. He, analyzing in detail the history of development of the Hellenides geosyncline in northern Greece, demonstrated that a geosyncline fundamentally consists of a couplet: the eugeosynclinal furrow called the Pindus Internides type and the miogeosynclinal furrow called the Ionian Externides type. The eugeosynclinal furrow is bounded by a eugeanticlinal ridge on the oceanic side, and separated by the miogeanticlinal ridge from the miogeosynclinal furrow. The sediments filling up both geosynclinal furrows are supplied from the eugeanticlinal ridge (Aubouin, 1965), first filling up the Pindus-type or eugeosynclinal furrow and then the Ionian-type or miogeosynclinal furrow (Fig. 2.18). In this interpretation, no supply of clastic sediments comes from the cratonic foreland or frontal stable region. Aubouin thus stressed a polarity not only in the sedimentary processes in geosynclines (Fig. 2.18), but also in magmatism, metamorphic processes and tectonic structures.

It is interesting to note that Kay did draw analogies of eugeosynclinal belts with modern volcanic arc systems. For example he noted comparisons made by H. H. Hess (professor of Princeton) between American serpentine belts and those of Caribbean islands (Dott, 1979).

**Fig. 2.17** Jean Aubouin (photo by H. Okada).

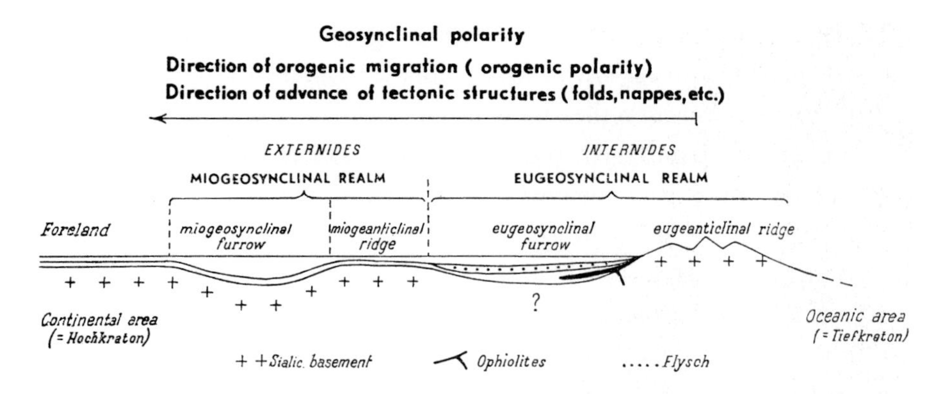


**Fig. 2.18** Aubouin's model of geosynclines (Aubouin, 1965) showing that the sediments first filled up the eugeosyncline and then, the miogeosyncline; the geosynclinal polarity is stressed.

Aubouin clearly indicated a correlation between past and present geosynclines. In the chapter 'The Geosynclinal Concept and the Present Epoch' in his book *Geosynclines* (Aubouin, 1965, p. 229), he states that the Sunda Islands and their surrounding seas in Southeast Asia are a present-day example of a geosyncline. In making this proposal he had taken account of new discoveries in marine geology. He reached this conclusion mainly with reference to the gravity surveys carried out by Vening Meinesz (1934, 1940, 1955) and other Dutch workers in the 1920s to 1940s. The structural patterns of the Sunda Islands in Indonesia, which Vening Meinesz and others clarified, show the two series of troughs: internal and external, and are thus comparable to the Hellenides of the Alpine chains.

C. L. Drake and his colleagues also recognized a present-day couplet of a eugeosyncline and miogeosyncline on the basis of the thickness distribution of sediments in their profile of the North American Atlantic coastal region (Drake *et al.*, 1959). Aubouin, however, criticized Drake *et al.*'s idea of present-day geosynclines, because it implied that all sediments bordering continents could thus be regarded as geosynclinal.

Subsequently, views favouring the existence of present-day geosynclines were boosted by Robert Sinclair Dietz (1914–1995) (Fig. 2.19), a professor of Arizona State University, one of the proposers of the Theory of Sea-floor Spreading. Dietz (1963, 1965) and his collaborators studied the geotectonic significance of a wedge-like profile of sediments across the continental shelf from off the North American Atlantic seacoast to the continental rise and demonstrated it to be an actualistic model of geosynclines (Dietz, 1963, 1965; Fig. 2.20).

**Fig. 2.19** Robert S. Dietz (photo by H. Okada).

According to Dietz, modern shelf sediments correspond to a miogeosyncline, and the wedge-shaped accumulation of sediments of the modern continental rise to a eugeosyncline. He further showed that the shelf topography does not make a depression like the miogeosynclinal model favoured by J. Aubouin, but inclines gently oceanwards and the shelf-sediments thus get clearly thicker oceanwards forming a wedge. Based on such a profile, Dietz and Holden (1966) coined a term 'miogeocline' for the shelf portion instead of miogeosyncline (*see* the shelf portion of Fig. 2.20A) and 'eugeocline' for the continental rise deposits.

Contrary to the classic geosynclinal concept, i.e. that extremely thick sediments characterize geosynclines that undergo orogenic movements and are deformed and uplifted, thus making an orogenic cycle (Dott, 1997), Dietz and others separated the hitherto accepted relationship between the geosyncline and orogenic movements. Later, Dott and Shaver (1974) gave an excellent interpretation of geosynclinal sedimentation in terms of the plate tectonics theory. Owing to its incorporation into the plate tectonic theory, the term 'geosyncline' has now become an obsolete word. This is because all the tectonic features explained before the 1960s and 1970s by geosynclinal theory as a geological paradigm have been reasonably explained by the tectonic activity caused by lithospheric plate movements: the plate tectonics theory or new global tectonics (Mitchell and Reading, 1969; Dewey and Bird, 1970; Dickinson, 1971).

Before leaving the subject of geosynclines, the Japanese contribution to the debate is worthy of mention. It was Teiichi Kobayashi, a professor of Tokyo University, who first applied the concept of geosynclines to the tectonic development of

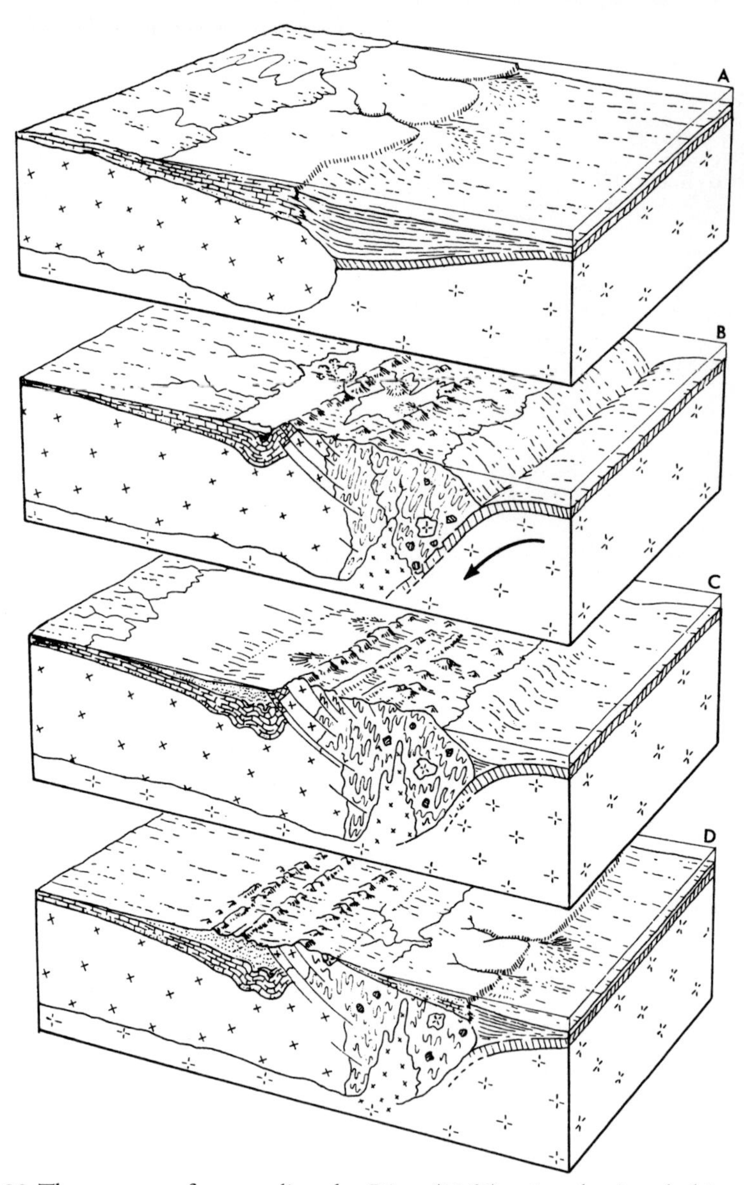


**Fig. 2.20** The concept of geosynclines by Dietz (1965) using the Appalachian region. It shows the tectonic development of the continental rise sediments (eugeocline) and the shelf sediments (miogeocline). A: initial stage of sedimentary accumulation in a miogeocline and a eugeocline (Latest Precambrian to Middle Ordovician), B: second stage showing the collapse of the continental rise due to the development of the trench system (orogenic movements caused by the start of subduction of the oceanic crust), causing plutonism and andesitic volcanism (Ordovician), C: third stage showing the end of orogenic movements, the end of plutonism and andesitic volcanism, and the deposition of molasse sediments in a miogeocline and filling of the trench (Late Palaeozoic), D: present-day aspect showing the tectonic stabilization of the present Appalachian region accompanied by the development of a continental terrace wedge as a new miogeosyncline and a continental rise as a new eugeosyncline. (Dietz, R.S. (1965), *Journal of Geology*, 73, Fig. 1 (p. 902). (*Reproduced by permission of the University of Chicago Press.*)

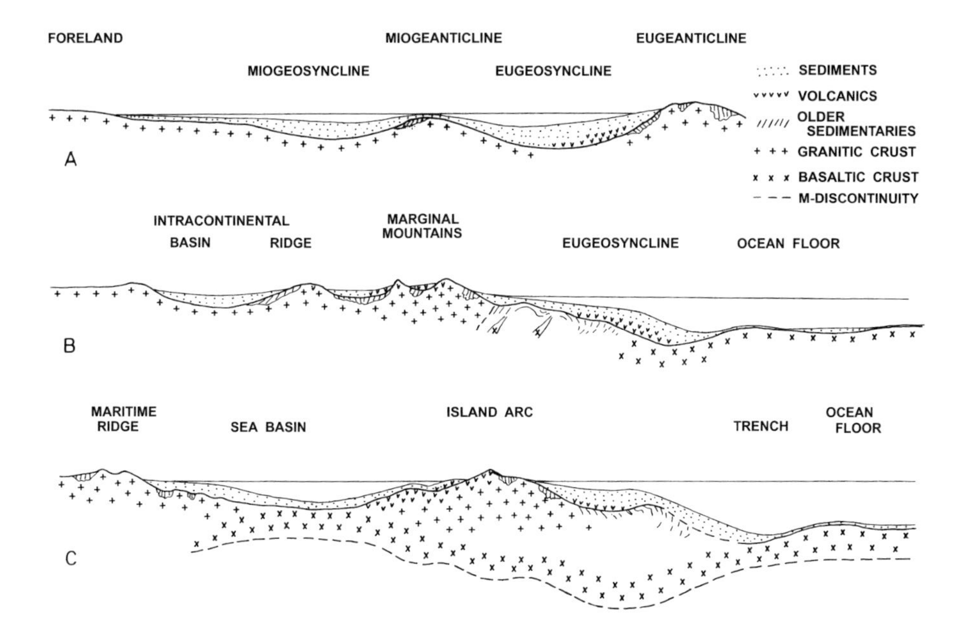


**Fig. 2.21** T. Matsumoto's model of geosynclines in profile (Matsumoto, 1967). A: European-type geosynclines, B: Mesozoic geosynclines in East Asia, C: Cainozoic geosynclines in the western margin of the Pacific. (*Reproduced with permission from Elsevier.*)

the Japanese Islands. Based on the Alpine orogenic interpretations of Kober and Stille, Kobayashi (1933, 1941, 1951) emphasized the relationship between the development of geosynclines and orogenic cycles in various stages of the geological evolution of the Japanese Islands. He recognized the following fundamental stages in the geological history of the Japanese Islands: the development of the Chichibu Geosyncline in the Permian to Jurassic; the Akiyoshi Orogenic Cycle in the Late Permian to Early Triassic; the development of the Shimanto Geosyncline from the Jurassic to Cretaceous; the Sakawa Orogenic Cycle of the Late Cretaceous; the development of the Nakamura and Yezo Geosynclines in the Late Cretaceous; and finally the Oyashima Orogenic Cycle of the Tertiary, mainly in the Miocene.

After Kobayashi's study, the Chichibu Geosyncline was called the Honshu Geosyncline and the Jurassic to Late Cretaceous Shimanto Geosyncline in the Outer Zone of Southwest Japan and the contemporaneous Hidaka Geosyncline in the axial belt of Hokkaido were researched in more detail. Subsequently, Tatsuro Matsumoto (1967), a professor of Kyushu University, showed that the couplet concept of miogeosyncline and eugeosyncline as established in Europe and North America could not be applied to the East Asian regions (Fig. 2.21).

It is natural that all of the orogenic and geohistorical developments of the Japanese Islands are now explained by the plate tectonics theory (Uyeda and Miyashiro, 1974 and Hashimoto, 1991).

# 3

# Development of the observation of strata

## 3.1 Sedimentary processes and the analysis of palaeo-environments

### 3.1.1 Lyell's observations of sedimentary structures

Charles Lyell, a founder, if not *the* founder, of modern geology, was an astute observer of the Earth's surface features and processes with the intention of using them for the interpretation of rocks (Leeder, 1998). This attitude led to his philosophy that the present is the key to understanding Earth's past. Lyell had a strong interest in sedimentary structures and tried to use them to interpret sedimentary rocks and depositional environments. Lyell pointed out in chapter 14 ('Sedimentary phenomena in deltaic sediments') of his book *Principles of Geology* (Lyell, 1830–1833) that strata are not always horizontal at initial sedimentation, but can be deposited with a considerable inclination. It suggests cross stratification, and he described a variety of cross bedding phenomena accompanied by outcrop sketches in which he illustrated herringbone cross stratification for the first time (Fig. 3.1, No. 33) but he had no idea of its significance. In addition, he discussed the relationship between cross stratification and ripple marks as well as that between the migration of asymmetrical ripple marks and current directions. He further considered that these sedimentary structures were formed in shallow water environments, thus identifying the relationship between sedimentary structures and depositional environments. Lyell paid considerable attention to the sedimentary structures of strata and realized that there was some relationship between sedimentary structures and sedimentary processes.

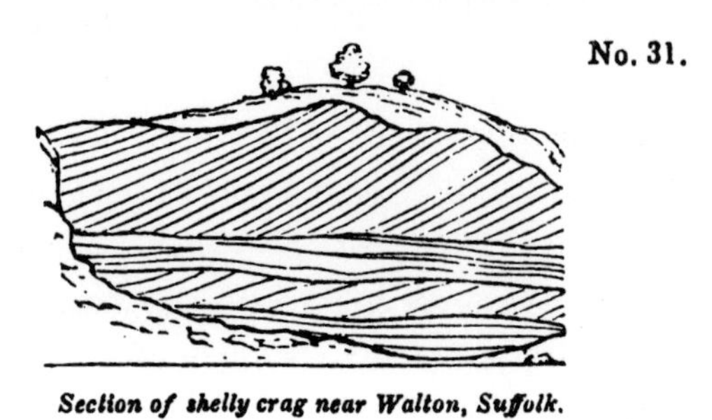


*Section of shelly crag near Walton, Suffolk.*

No. 32.

a

b

c

c

*Section at the lighthouse near Happisborough. Height sixteen feet.*

**a, Pebbles of chalk flint, and of rolled pieces of white chalk.**
**b, Loam overlying a.** **c, c, Blue and brown clay.**

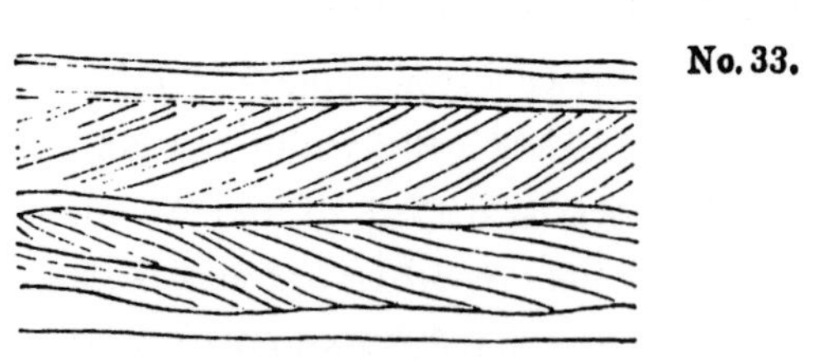


*Section of part of Little Cat cliff, composed of quartzose sand, showing the inclination of the layers in opposite directions.*

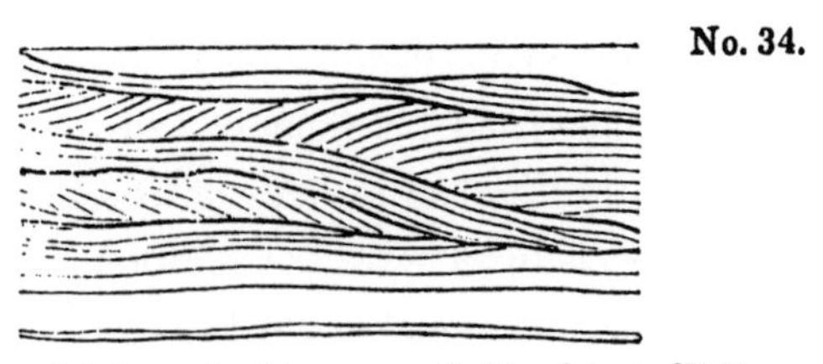


*Lamination of shelly sand and loam, near the Signal-house, Walton. Vertical height four feet.*

Fig. 3.1 Lyell's field sketches of cross stratification (Leeder, 1998). Lyell sketched the cross-stratification from the Plio-Pleistocene in East Anglia (Lyell, 1830–1833, vol. 3, pp. 174–175; figs. 31–34). Figure in No. 31 is about 6.1 m high, in No. 32 about 4.9 m high, in No. 33 about 1.8 m high, and in No. 34 about 1.2 m high. No. 33 is the first illustration of the herring-bone cross-stratification. (*Reproduced by permission of the Geological Society of London.*)

For such acute observations of sedimentary phenomena, Leeder (1998) exalted Charles Lyell as the true originator of sedimentology, rather than H. C. Sorby. It seems, however, that Leeder gave Lyell too much credit. Though he called attention to many sedimentary structures, he had only rudimentary ideas about their origin. Moreover, these were not at all Lyell's main focus. It should be recorded here that Sorby was indeed the father of sedimentology as discussed below and on p. 84.

At about the same time, James Hall, famous for the proposal of the geosynclinal concept, also recognized evidence of current flows in sediments produced at the time of their sedimentation when he studied Appalachian sedimentary rocks (Hall, 1843).

### 3.1.2 Sorby's recognition of sedimentary structures and his interpretation of sedimentary processes

It was Henry Clifton Sorby (1826–1908) (Fig. 3.2), often said to be the father of sedimentology, who quantitatively studied for the first time the origin and mechanism of the formation of sedimentary structures. Based on his observations of the sandstones of the Carboniferous Coal Measures in central England, he noted the relationship between bed forms and current velocities, or, according to present terminology, the relationship between primary sedimentary structures and flow regimes (Sorby, 1852). In his study, Sorby noted changes of current velocities and the size of ripples marks showing that beyond a certain current speed, or the threshold value, cross bedding or false bedding with an inclination of about 30° was produced. When the current becomes more rapid, ripples are formed on the bottom surface and sand particles are carried forward and drifted ripples appear (Fig. 3.3A, B). Furthermore, the ripples formed indicate regular orientations suggesting currents with definite directions. Sorby had in mind an idea that sedimentary structures should be treated on the basis of the relationship between field occurrences and experiments.

In such a manner, Sorby had the strong opinion that observations and experiments should always go together. This idea was clearly expressed in his presidential address to the Geological Society of London entitled 'On the application of quantitative methods to the study of the structure and history of rocks', which was the summary of his life-work and is valued as an excellent and masterly scientific account (Sorby, 1908). In the opening sentences of this paper, he stated that 'My object is to apply experimental physics to the study of rocks', a current theme even today. The topics he discussed in his paper include the

**Fig. 3.2** Henry Clifton Sorby (Judd, 1908).

relationship between sizes of clastic grains and current velocities in hydraulic experiments, angles of rest of sand grains and pebbles, velocities required to transport particles, ripple-drift, characteristics of fine-grained sediments such as clay and chalk, diagenesis and pressure solution. In particular, a table showing sedimentation rates of mud particles after two hours and eight days, respectively, in his experiments is impressive. In his paper, the term 'ripple mark' was introduced.

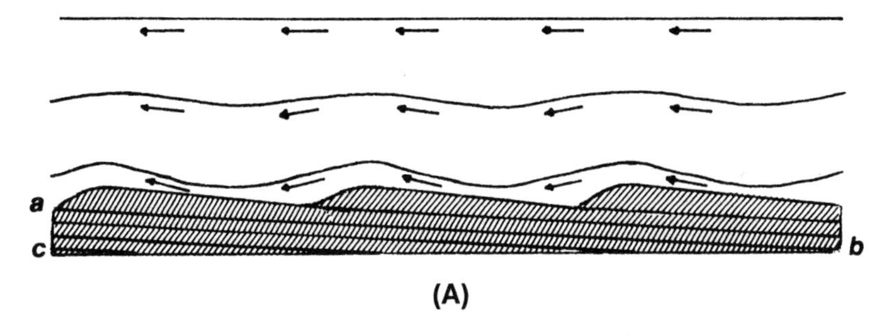


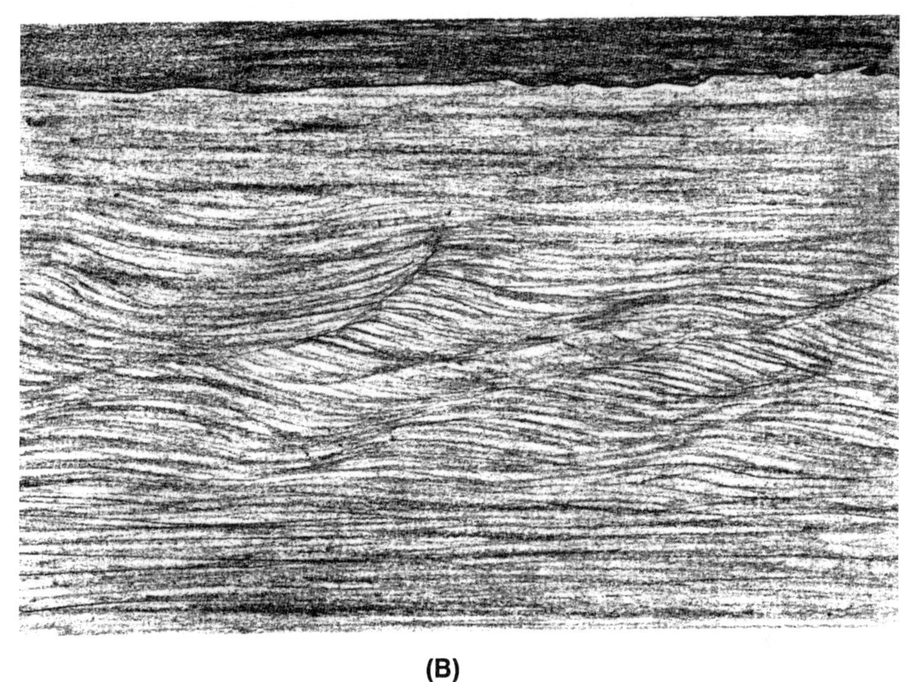


**Fig. 3.3** Sorby's sketch of ripple-drift cross-stratification. (A) structure of ripple-drift cross-stratification (Sorby, 1859, fig. 3 in Summerson, 1976, p. 31). a-c: a bed of ripple-drift, a-b: ripple-drift-bands. (B) sketch of a ripple-drift layer from the Green Slate, Langdale, Yorkshire, England (Sorby, 1908, pl. xvi). Inferred ripple wavelength is about 9 cm.

### 3.1.3 Sedimentary structures: indicators of younging directions in strata and sedimentary environments

According to R. H. Dott, Jr. (2003a), a professor of the University of Wisconsin, the value of sedimentary structures in field studies was first recognized as early as in the mid-nineteenth century when Patrick Ganly (1856) showed the correct way up of a stratigraphic succession by using crossbedding structures. He indicated the

**Fig. 3.4** Edward Battersby Bailey (Stubblefield, 1965).

original way-up direction in folded Silurian sandstones in the Dingle Peninsula in southwesternmost Ireland. After Ganly, more works using sedimentary structures as way-up indicators were made by E. B. Bailey and many other geologists, as summarized by R. H. Dott, Jr. (2003a).

The sedimentary structures described by H. C. Sorby were put to practical use as a fundamental contribution to studies in the field by Edward B. Bailey (1881–1965) (Fig. 3.4) (Pettijohn, 1975), a geologist who became the Director of the British Geological Survey, and he showed the value of such studies when carrying out geological surveys and reconstructing sedimentary environments. Bailey (1930, 1936) noted that there are two principal types of internal structure in beds: cross bedding and graded bedding (Fig. 3.5) and that they are characteristic of two contrasting depositional facies: graded bedding is a characteristic facies of deep water deposition, while crossbedding has a rather complicated origin, but also is useful for suggesting depositional environments. According to R. H. Dott, Jr., it was two young Wisconsin geologists Olaf Rove and Sherwood Buckstaff who examined deformed strata in the Scottish Highlands with the

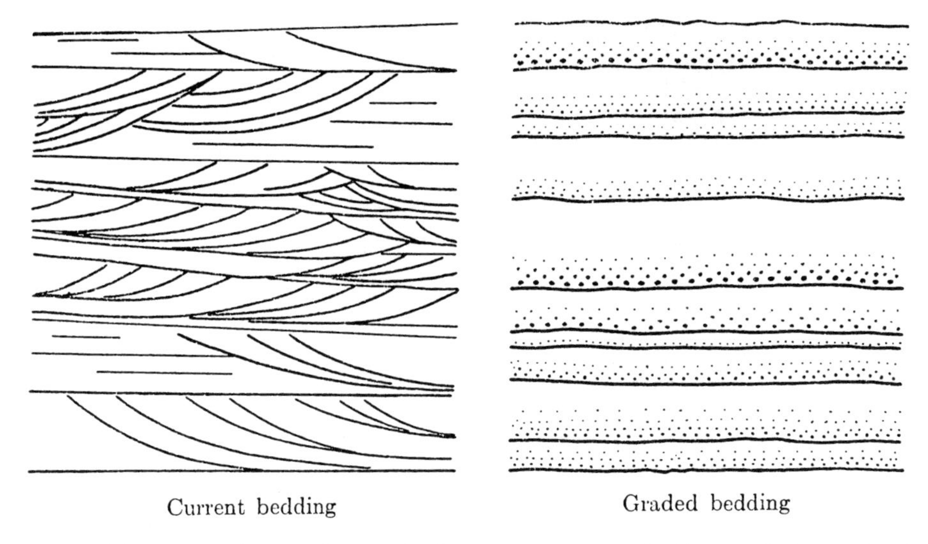


**Fig. 3.5** Two principal types of internal sedimentary structures (Bailey, 1936). (*Reproduced by permission of The Geological Society, London.*)

Norwegian geologist Thorolf Vogt, in 1924 and later introduced E. B. Bailey to the importance of sedimentary structures for determining the original way-up attitude of strata (Dott, 2001).

Bailey (1930) used cross bedding to identify the younging direction of strata in a study of the tectonically complicated metamorphic zone of the Ballachulish Belt in the Highlands of Scotland. The structure was significant because the upper surfaces of cross-bedded layers show a concave-upward form and the uppermost ends of cross-bedded layers meet the overlying strata at large angles while the opposite, lower ends converge with the underlying beds at very small angles (Fig. 3.5 *left*). His observation had been confirmed and enthusiastically supported by Thorolf Vogt, a professor of Trondheim University, and T. L. Tanton of the Canadian Geological Survey on a field excursion to the area in 1929 (Vogt, 1930; Tanton, 1930). Thus, it became clear that sedimentary structures are useful not only for identifying the original order of succession of strata but also in analysing geological structures in tectonically complicated regions.

Bailey (1930) also pointed out that the sedimentary structure of graded bedding is an equally important indicator of the order of succession of strata (Fig. 3.5 *right*). Graded bedding is a sedimentary phenomenon in which grain sizes at the base of a unit bed are coarsest, gradually get smaller upwards with the finest at the top. Bailey observed that graded bedding is not usually associated with cross bedding.

**Fig. 3.6** Graded bedding from the Cambrian Barmouth Formation east of Barmouth, North Wales (photo by H. Okada in 1964).

Bailey presented good examples of cross bedding in the Devonian Old Red Sandstone, Carboniferous sandstone beds and the Permo-Triassic New Red Sandstone, but noted that almost no examples of graded bedding occur in these strata. Conversely, he suggested that graded bedding characterizes the Lower Palaeozoic of Wales. As type localities of graded bedding, he pointed to the exposures of the Cambrian Barmouth Grits below Harlech Castle and at Barmouth for coarse-grained graded beds (Fig. 3.6), and to those of the lower Llandoverian Aberystwyth Grits of the Lower Silurian on the seacoast south of Aberystwyth, Central Wales (Fig. 3.7).

The genetic environments of cross bedding and graded bedding were considered to be as follows: cross bedding was formed by winds and currents in shallow waters, whereas graded bedding must have been produced by coarse-grained sediments transported intermittently from shallow coastal areas into quiet deep-sea environments by currents initiated by earthquakes or floods. Bailey (1936) illustrated for the first time that the sedimentary structures of current bedding (cross bedding) and graded bedding are not related, emphasizing that graded bedding characterizes flysch sandstones in geosynclinal sediments.

**Fig. 3.7** The lower Silurian Aberystwyth Grits Formation at Monks Cave south of Aberystwyth, central Wales (photo by H. Okada in 1964).

An additional feature of Bailey's publications was that he carefully defined his usage of technical nomenclature and if no technical term was known to him, he did not hesitate to coin one. Thus, in his 1934 paper he introduced the verb 'to young' to reflect that one part of a sequence was younger, that is later, than another (Stubblefield, 1965).

After Bailey's works on graded bedding and current bedding as reliable top and bottom criteria, the study of sedimentary structures quickly became, and was widely accepted as, the most effective method for determining the younging direction of strata; a concept that was to be refined and elaborated by R. R. Shrock (1948).

## 3.2 Palaeocurrent analysis

The fact that currents leave an imprint as sedimentary structures in sediments was first pointed out by Charles Lyell (1830) in England and James Hall in America. H. C. Sorby emphasized the importance of systematic observations of orientations of cross bedding. Later, Bailey's (1930) observation that such sedimentary structures as cross bedding and graded bedding played essential roles in characterizing sedimentary facies and analysing regional tectonic developments was corroborated by the case studies of R. Brinkmann (1933), Hans Cloos (1938), C. I. Migliorini (1943) and others in many areas of Europe and North America.

From such beginnings, growing interest in the evidence of current directions preserved in sediments and sedimentary rocks stimulated the developments of a variety of methods: for example, the analyses of sole marks on the lower surface of sedimentary beds, grain orientation or fabric of sands and gravels, downcurrent size reduction of particles and the preferred orientations of fossil shells and plant fragments. Among them, studies of cross bedding were always regarded as decisive for analysing current directions of sediments and sedimentary rocks in any geological period. The studies of F. J. Pettijohn (1962) can be regarded as representative of such researches. It should be added that Joseph Barrell (1869–1919) was an American pioneer for understanding sedimentation in terms of tectonics, deposition and climatic conditions in sedimentary basins (Barrell, 1906) (Dott, 2003a,b).

Palaeocurrent analyses carried out by collecting statistical data regarding surface structures (for example, current directions obtained from the foresets of cross-bedding) and linear structures (for example, ripple marks and erosional structures) have become an indispensable method for the study of modern and ancient sediments. In this respect, the term 'palaeocurrent' seems to have been used first by Pettijohn (1957b). Such directional data can be treated as vector values for vectoral analysis, as can many scalar properties which reveal systematic areal downcurrent changes in numerical values of grain size, changes of roundness and changes in thickness of the strata.

In this latter regard, H. Sternberg (1875) quantitatively observed for the first time the processes of downcurrent decrease in gravel sizes in an actual river. Measuring the maximum and average sizes of gravels over a distance of about 250 km in the Rhine River from Basel downstream, he concluded that the size and weight of gravels decrease exponentially with the distance of transportation, and formulated the following equation:

$$W = W_0 e^{-as}$$

where $W$ is the weight of a gravel after transport for the distance of $s$, $W_0$ is the initial weight of a gravel, and $a$ is the coefficient of size reduction.

The relationship expressed by this formula is called 'Sternberg's Law'. Since then there have been many examples of quantitative studies such as the recognition of flow directions using changes of gravel size (Bluck, 1965) as well as in the construction of palaeogeography (Yeakel, 1962). It should be noted, however, that the rate of particle size reduction downstream is not in all cases an exponential function, and Sternberg's Law cannot be applied to particles smaller than granules. The rate of size decrease is controlled by the size of grains, gravels larger than pebbles being reduced more quickly than sands. E. Yatsu (1954–55, 1955) found an abrupt change in the rate of size diminution between pebbles and sands in his study of river deposits in Honshu, Japan.

## 3.3 Basin analysis and megasedimentology

Dispersal patterns of sedimentary bodies are important to understand the shape and size of a sedimentary basin. In a landmark study, Edward Hull (1862) published the first geological map to show the thickness and distribution of sediments using isodiametric contours in his work on the clastic sediments and limestones of the Carboniferous System in part of Great Britain.

Henry Clifton Sorby in England was the first to pay attention to the orientation of cross stratifications, but he did not publish his measured data. Many years later Francis John Pettijohn (1904–2000) (Fig. 3.8), a professor of Johns Hopkins University, became convinced that the study of palaeocurrents provided the most effective method for analysing the provenances of sediments as well as for the reconstruction of sedimentary basins. This was the time (1952) when Pettijohn moved to Johns Hopkins University from Chicago University. At that time, the Appalachian geosyncline west of Baltimore was being studied by Marshall Kay who postulated tectonic uplands and volcanic islands within and to the east of the geosyncline as sedimentary sources. Pettijohn, based on his own studies of sandstone composition, questioned Kay's opinion.

In order to find a solution to this enigma, Pettijohn sent Paul E. Potter, his last student at Chicago University, and Bern E. Pelletier and Earle F. McBride, his first students at Johns Hopkins, to the Appalachian Palaeozoic to make a systematic collection of the data on palaeocurrent orientations and the mineral composition of the clastics. As a result, the Appalachian Palaeozoic sands indicated derivation from older found to the west sedimentary and metamorphic rocks, not from eastern volcanic rocks, and, thus, a new vision was developed of the provenance of the Appalachian Palaeozoic sediments (Potter and Olson, 1954; Pelletier, 1958; Pettijohn, 1962).

**Fig. 3.8** Francis J. Pettijohn (photo by H. Okada).

As stated above, the work of determining the current systems in sediment transportation is defined as palaeocurrent analysis. Palaeocurrent analysis is based on *all* available primary sedimentary structures including cross bedding. In addition to palaeocurrent analysis, the lithofacies of sediments, sedimentary sequences, three-dimensional features and tectonic setting are all significant in analyzing the entire nature of a sedimentary basin. Using all these methods Pettijohn, in collaboration with one of his students, Paul Edwin Potter (Fig. 3.9) of Indiana University, established a new field of sedimentology, which they called 'Basin Analysis' (Potter and Pettijohn, 1963). This quickly became the fundamental approach for the evaluation of sedimentary basins. Philip A. Allen and John R. Allen (1990) define *basin analysis* as the integrated study of sedimentary basins as geodynamic entities.

**Fig. 3.9** Paul Edwin Potter (Pettijohn, 1984).
(*Reproduced by permission of The University of Chicago Press.*)

The integrated study of the sedimentology of sedimentary basins over vast regions, for example ancient and modern continental margins, or of entire fold belts, such as the whole Alpine or Appalachian or Andean mountain chains, regarding their shape, palaeocurrent systems, lithology and depositional processes of sediments and tectonic environments, is called *megasedimentology* (Potter, 1978; Friedman and Sanders, 1978).

The fundamental principle of basin analysis is to characterize the general features of sedimentary basins on the basis of analyses of grain fabric in unit beds, current analysis using sedimentary structures, and the recognition of the three-dimensional geometry of sediments, sedimentary units and lithofacies (Miall, 1984).

Basin analysis is readily applicable to on-land sedimentary basins and shallow-sea areas where the marginal facies of basins are well preserved and three-dimensional features of sediments are easily observable. In contrast, it is generally difficult to observe the entire character of sedimentary basins in deep-sea environments. This is because ancient flysch sedimentary basins, for example, show that many sedimentary facies are in fault/thrust contact with each other, and modern examples are too deep to determine the entire form of the sedimentary basin.

Good examples of basin analysis in the case of on-land and shallow-sea environments are represented by the studies of the Mississippian Pocono Formation in the central Appalachians (Pelletier, 1958) (Fig. 3.10) and the Silurian Tuscarora Formation (Yeakel, 1962); in the case of flysch basins by the study of the Cretaceous to Palaeogene flysch in the Carpathian mountains (Ksiazkiewicz, 1960) (Fig. 3.11); and in the case of modern deep-sea environments by the study of the Hispaniola–Caicos Basin in the Caribbean Sea (Bennetts and Pilkey, 1976) (Fig. 3.12).

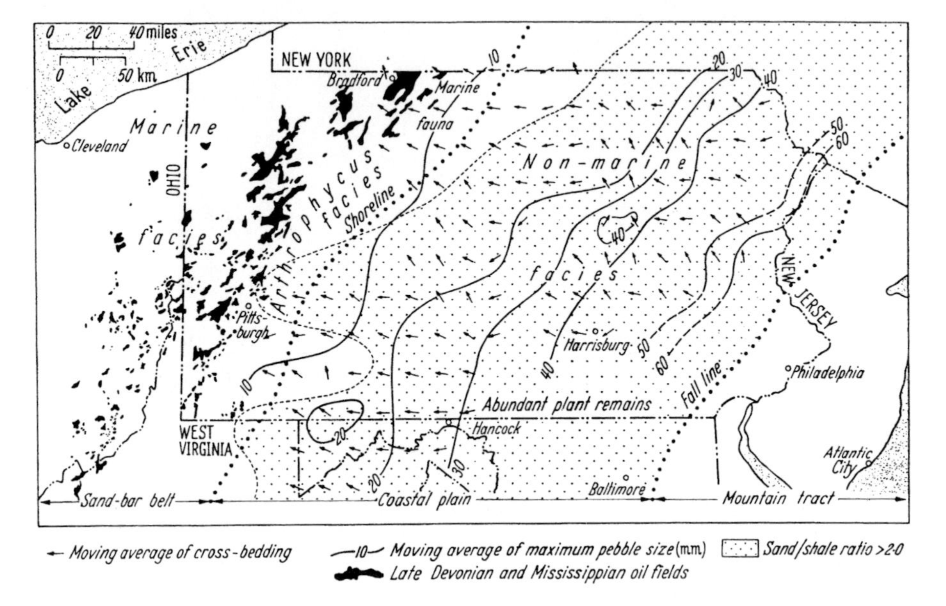


**Fig. 3.10** Basin analysis of the Mississippian Pocono Formation in the central Appalachian (Pelletier, 1958). (*Reproduced by permission of the Geological Society of America.*)

***Opposite*** **Fig. 3.11** Basin analysis of the Cretaceous–Palaeogene Carpathian flysch south of Krakow, Poland (Ksiazkiewicz, 1960). (A) *Valanginian Stage*: 1. flysch facies, 2. limestone facies, 3. approximate position of the shoreline. (B) *late Senonian Stage*: 1. epicontinental marine deposits in the foreland, 2. variegated marls, 3. grey marls, 4. Szydlowiec sandstones, 5. Lower Istebna sandstones and conglomerates, 6. Inoceramian beds, 7. Baculite marls, 8. Puchow marls, 9. Jarmuta sandstones, 10. Bachowice red limestone. (C) *middle Eocene Stage*: 1. fine-grained beds of the Beloveza and Hieroglyphic beds, 2. variegated marls, 3. grey Popiele marls, 4.coarse-grained Pasierbiec sandstones, 5. grey Loncko marls, 6. conglomerates and nummulitic limestones: a – a: present northern border of the Carpathian Range, b – b: Pieniny Klippen Belt, s – s: northern margin of the Silesian Nappe, d: northern margin of the Dukla Zone, m: northern margin of the Magura Nappe. Arrows indicate main directions of transport. Note the appearance of uplifts in the central part of the basin, as shown in B and C.

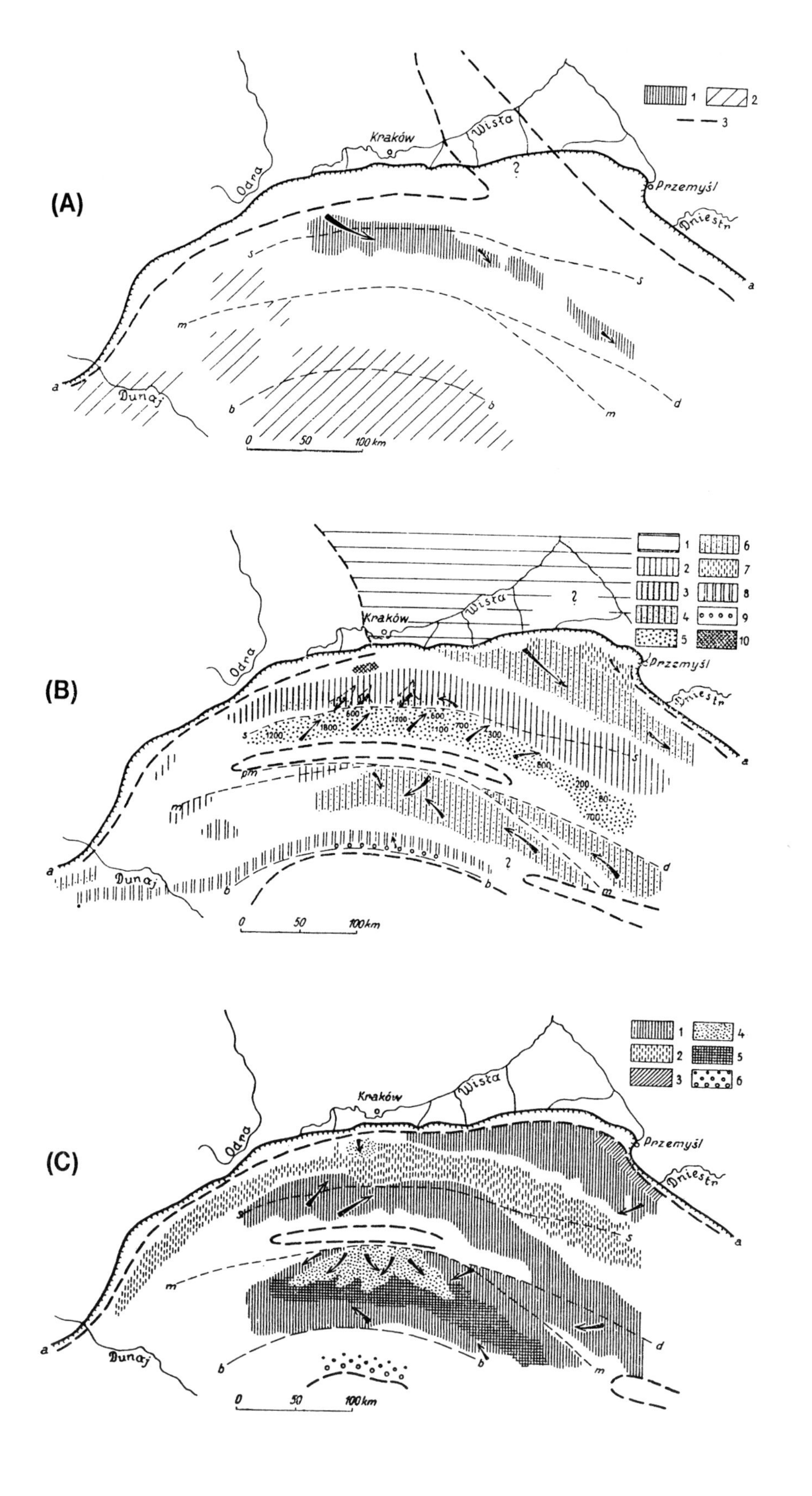

(A)
Kraków
Wisła
Odra
Przemyśl
Dniestr
Dunaj
0 50 100 km
(B)
Kraków
Wisła
Odra
Przemyśl
Dniestr
Dunaj
0 50 100 km
(C)
Kraków
Wisła
Odra
Przemyśl
Dniestr
Dunaj
0 50 100 km

**Fig. 3.12** Basin analysis of the Hispaniola–Caicos Basin (Bennetts and Pilkey, 1976) (a) Index map of the Hispaniola–Caicos Basin, (b) Bathymetric map of the Hispaniola–Caicos Basin, (c) Physiographic provinces on the Hispaniola–Caicos Basin floor, (d) Isopach map of the three individual turbidites, (e) Possible flow directions of the three turbidites based on turbidite thickness, (f) Isopach maps of individual Bouma divisions. (*Reproduced by permission of The Geological Society of America.*)

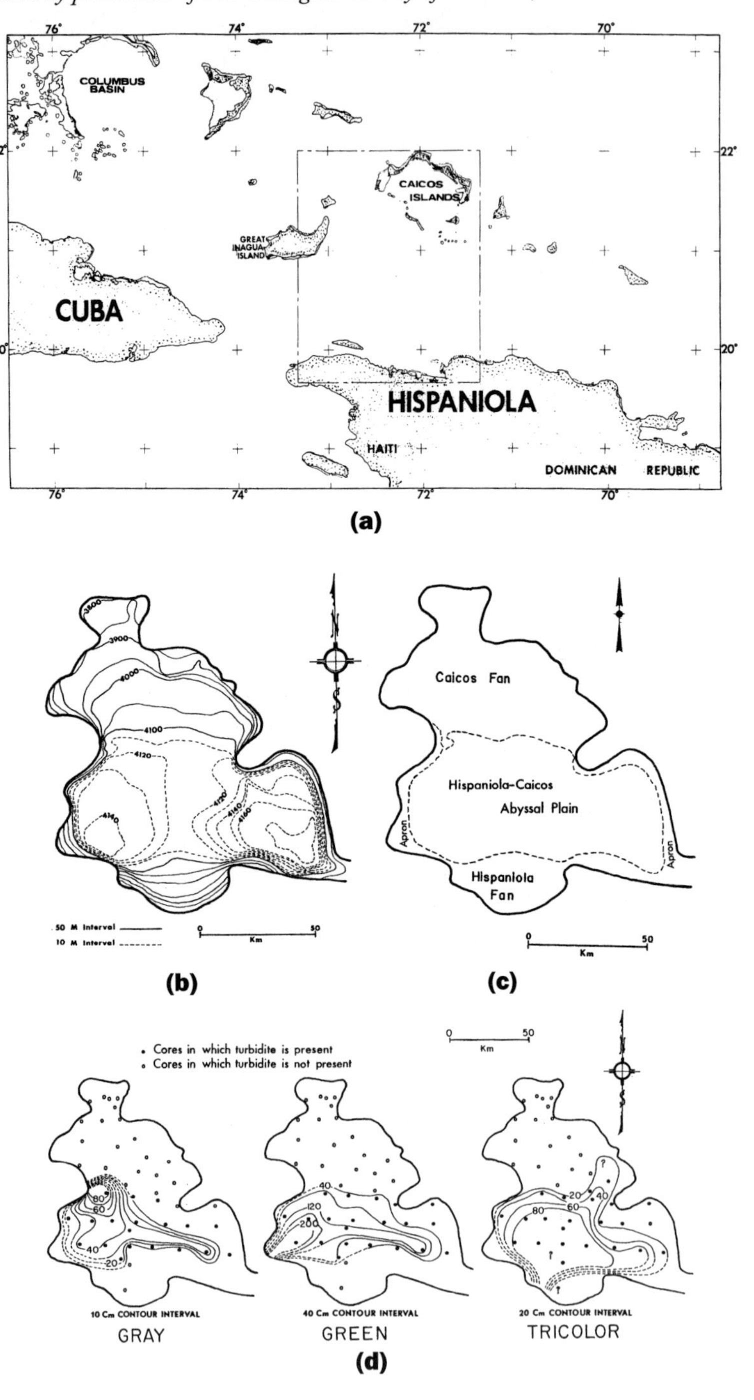

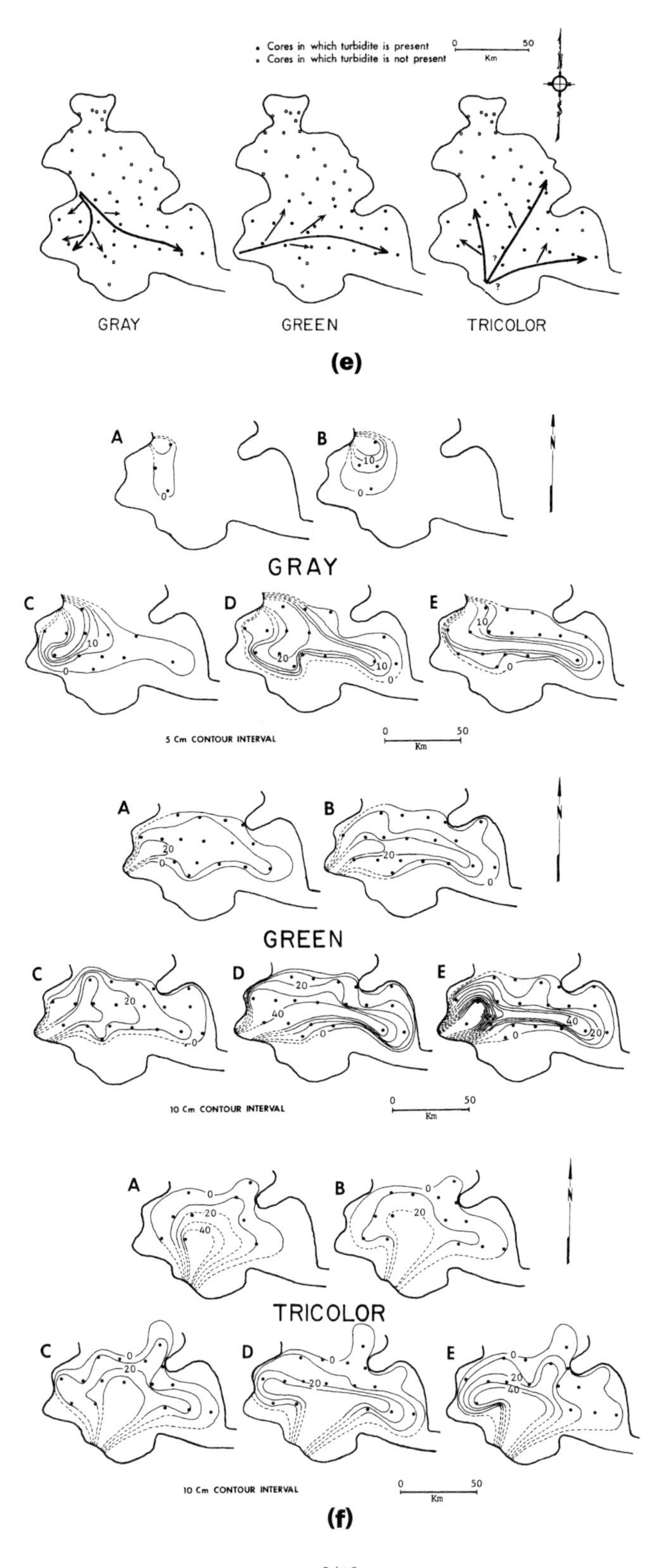

Cores in which turbidite is present
Cores in which turbidite is not present
0
50
Km
GRAY
GREEN
TRICOLOR
(e)
A
B
GRAY
C
D
E
5 Cm CONTOUR INTERVAL
GREEN
10 Cm CONTOUR INTERVAL
TRICOLOR
(f)

The Hispaniola–Caicos Basin, situated in the Caribbean Sea between the Caicos Islands and Hispaniola, is on average about 4 km deep and is 9500 $km^2$ in area. The deepest part of the basin, in the southeast, reaches a depth of 4160 m (Fig. 3.12b). Three layers of turbidite are found in the deeper part (> 4100 m) of the Hispaniola–Caicos abyssal plain (Fig. 3.12c). They are colour characterized into an upper Gray Turbidite, a middle Green Turbidite, and a lower Tricolor Turbidite. The average thickness of each layer is 33 cm, 97 cm, and 60 cm, respectively; the maximum grain size is + 1.41 phi, – 0.53 phi, and + 0.77 phi, respectively. The phi (ø) number (Krumbein, 1934) used here refers to a logarithmic transformation of the Wentworth grade scale (*see* Chapter 6.1.3). Each turbidite layer is traceable for a distance of about 100 km (Fig. 3.12d). Each division of the Bouma sequence (*see* below, 3.4.6) is well observed over a wide area (Fig. 3.12f), and the behaviour of the turbidity currents can be determined (Fig. 3.12e).

## 3.4 Advent of the turbidity current concept

### 3.4.1 Density current and turbidity current

It had long been observed that from time to time turbid waters from the River Rhône enter Lake Geneva (lac Léman), Switzerland, and submerge under the clear lake water to reach the deepest parts of the lake. F. A. Forel (1887) noticed the existence of canyons cut into the delta where the Rhône entered the lake. He ascribed their origin to erosion of the lake floor by the bottom sweeping currents of turbid water, and pointed out that the turbid water was not only lower in temperature than the surrounding lake water, but also denser and heavier due to suspended sediment. This was the start of the density current concept, that is, a water mass of higher specific gravity due to lower temperature, salinity and/or suspended sediment than the surrounding water will sink and thus create a density current.

It was, however, R. A. Daly (1936), who suggested for the first time the importance of density currents as a geological phenomenon. He considered that deep-sea submarine canyons could have been eroded by sediment-laden currents (density currents) pouring down the continental slope. This hypothesis was supported by Douglas W. Johnson (1939), and he first proposed the term 'turbidity current' for such currents in order to distinguish them from high density currents merely due to lower temperature and higher salinity. The water/sediment body flows down slope because of gravity.

### 3.4.2 Experimental verification of the turbidity current

The sedimentary structure called graded bedding by E. B. Bailey in 1930 was considered to be produced in a deep-sea environment by some sort of bottom current. R. A. Daly (1936) proposed the concept of density current as an erosive rapid flow on the deep continental margin, but the nature and behaviour of the density current had not been well understood. Taking a clue from this idea, Philip Henry Kuenen (1902–1976) (Fig. 3.13) set out to confirm experimentally the actual nature of a density current (Kuenen, 1937). However, it is proper to mention that the existence of sediment- transporting bottom currents had been recognized in Lake Mead behind the Hoover Dam, Nevada (Grover and Howard, 1937), and their sediments were studied by Gould (1951). Engineers had become concerned about the rapid sediment-filling of Lake Mead by these 'new' currents and H. S. Bell (1942) at California Technical University was contracted to do

**Fig. 3.13** Philip H. Kuenen (photo by H. Okada).

experiments with turbidity currents in the 1940s – before P. H. Kuenen (*see* Dott, 1978). The laboratory experiments carried out by Kuenen (1950a) used a small inclined water tank divided by a partition wall into two parts. When the partition was removed the turbid high-density water of the upper part flowed quickly down the bottom slope under the clean water part of the second and a graded structure was created when the turbid water sedimented out (Kuenen, 1950a).

He, at the time a professor of Groningen University in the Netherlands, then postulated, in collaboration with C. I. Migliorini, who was studying the Oligocene flysch in the northern Apennine Mountains, that unit sandstone beds of flysch sediments exhibiting graded bedding were the deposits of turbidity currents (Kuenen and Migliorini, 1950). This was the first recognition that ancient large-scale flysch deposits are the product of sedimentation by turbidity currents. Kuenen called the graded sediments deposited by turbidity currents *turbidites* (Kuenen, 1957).

After Kuenen's pioneering work, hydraulic experiments on turbidity currents were carried out by Gerard V. Middleton (Fig. 3.14), a professor of McMaster

**Fig. 3.14** Gerard V. Middleton.

University in Hamilton, Canada. From his experiments, the hydraulic behaviour of turbidity currents became much clearer, in particular, the density difference within a turbidity current surge which created a head, neck, body and tail in the current (Middleton, 1966a, b, c; 1967). Further contributions to the detailed analyses of the hydrodynamics of turbidity currents were made by, among others, J. R. L. Allen (1971), Komar (1971) and Edwards (1973).

### 3.4.3 Sedimentary structures produced by turbidity currents

Since Kuenen and Migliorini (1950) clarified the mechanism of graded bedding characteristics of turbidites, more information on the mechanism for producing internal sedimentary structures of turbidites has been determined through experiments as mentioned above. At the same time, studies of sedimentary structures produced by turbidity currents and recorded either on the top or the bottom of beds were carried out through field observations as well as in laboratory experiments: for example, Rich (1951), Dzulynski and Walton (1963, 1965); Middleton (1965); Sanders (1963); J. R. L. Allen (1971).

Pettijohn (1975) classified the structures of sedimentary rocks into mechanical ('primary'), chemical ('secondary') and organic structures. Mechanical or primary sedimentary structures include bedding, bedding internal structures (cross bedding and graded bedding are included), bedding-plane markings (on sole and on surface) and deformed bedding. The bedding-plane markings (on the sole) are further divided into scour or current marks and tool marks, which are important structures generally found on the underside of a turbidite bed. Thus, they are called sole marks as a descriptive term. Scour or current marks are the structures eroded into finer-grained sediments by a turbulent current, as represented by flute casts. Tool marks are the structures produced on soft muddy sediment surfaces by moving or stationary objects (tools) during deposition. Several kinds of tool marks such as groove marks, chevron marks, prod marks, bounce marks and brush marks are recognized.

These structures are all important not only for determining flow conditions and current directions at the time of sediment deposition, but also for elucidating the form and palaeobathymetry of the depositional basins. As major contributions to the study of sole marks, Dzulynski *et al.* (1959), Dzulynski and Walton (1965), Ricci Lucchi (1995) and Dzulynski (2001) are important.

### 3.4.4 Observations of actual turbidity currents

It was claimed by extrapolation that turbidity currents, as observed in laboratories and Lake Mead, Nevada, must have deposited ancient graded beds. Convincing proof, however, could only come through the discovery of a turbidity current and associated deposit in the oceans of the present. This proof was provided quickly through a review of the records of the Grand Banks earthquake of 18 November 1929 (Heezen and Ewing, 1952). The records indicated that only a turbidity current generated by the earthquake could have sequentially broken the trans-Atlantic submarine telegraph cables in the vicinity of the Banks.

Bruce C. Heezen and Maurice Ewing of the Lamont-Doherty Geological Observatory checked the records of the precise time of each break of the trans-Atlantic cables on the continental slope south of Newfoundland. They found that while the cables on the continental shelf were not broken, those in the epicentre area on the upper continental slope were broken simultaneously, and those away, down slope, from the epicentre area were broken in sequence (Fig. 3.15a, b). Using these data Heezen and Ewing obtained the time of arrival of the turbidity current at successive sites, and thus could calculate the velocity of the current (Fig. 3.15b): a velocity of 55 knots (= 99 km/h = 28 m/s) at the most proximal site on the slope, then a gradually decreasing speed, and a velocity of 12 knots (= 21.6 km/h = 6 m/s) at the most distal site on the bottom slope of less than 1/1500. The turbidity current flowed for the distance of 648 km to the depth of about 5 km in 13 hours and 17 minutes.

The hydraulic features of the Grand Banks turbidity current were further re-examined by Kuenen (1952), Shepard (1967), and Hsü (1989). According to Hsü (1989), the Grand Banks turbidity current should be regarded as a sediment gravity flow in the present terminology. Nevertheless, the Grand Banks event of 1929 provided evidence of natural turbidity currents that are capable of travelling a great distance.

As to the significance of turbidity currents: they transport large amounts of clastic sediments into the deep ocean, they bring abundant organic materials into the nutrient-starved deep-sea floor, and they suggest a source area with tectonically unstable conditions. The trigger mechanisms of turbidity currents include earthquakes, storms such as typhoons and hurricanes, high inflows of muddy river currents and sediment collapse at the outer continental shelf margin.

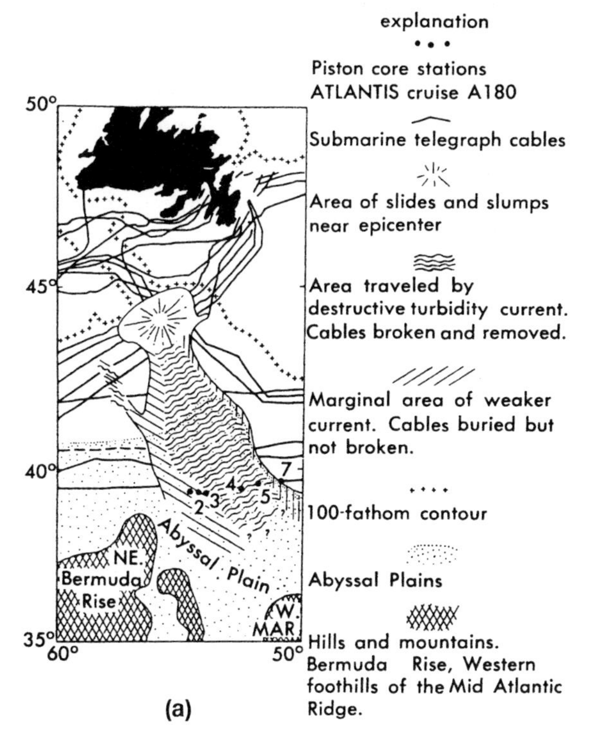


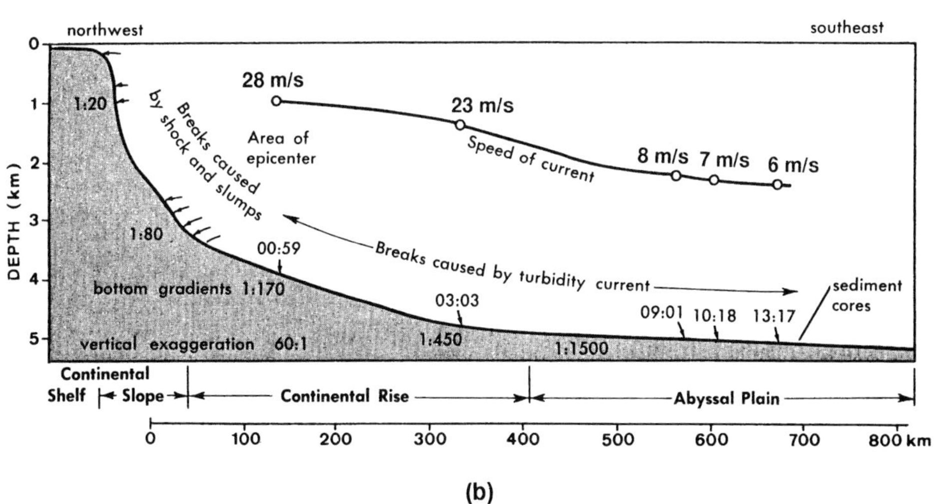


**Fig. 3.15** (a) Area of the Grand Banks earthquake of eighteenth November, 1929 and (b) profile of the submarine topography south of the Grand Banks showing the progress of the turbidity current (modified from Heezen, 1960). (*Reproduced by permission of the Mc-Graw Hill Companies.*)

### 3.4.5 The gravity-flow concept

A recent trend has been to recognize that turbidity currents are one of a number of mass movements that include landslides and slumps. A factor controlling the nature of such flow processes includes the rate of mixing of liquids and solid particles. According to M.A. Hampton (1972), when the proportion of solid particles increases under the influence of gravity, liquid flows change into particle flows. Middleton and Hampton (1973) proposed the term *sediment gravity flow* for flows of mixtures of sediments and water that move down a slope under the effect of gravity, be they in subaerial or subaqueous environments.

When solid particles in sediment–water mixtures are affected by gravity and the whole mixed liquid flows downslope, the following two mechanisms for continuing flows are recognized: (1) the shear stress given by the downslope gravity component on the mixed fluid exceeds the frictional resistance to flow, and (2) the grains are held apart in the liquid by such supporting mechanisms as (i) turbulence, (ii) buoyancy, (iii) dispersive pressure by grain collision, (iv) trapped or escaping pore fluids, (v) cohesive strength, and (vi) frictional strength. The sediment gravity flows are classified by sediment-supporting mechanisms into the following four flow types: *turbidity current*, in which the main mechanism supporting sediments is turbulence, *fluidized sediment flow*, in which grains

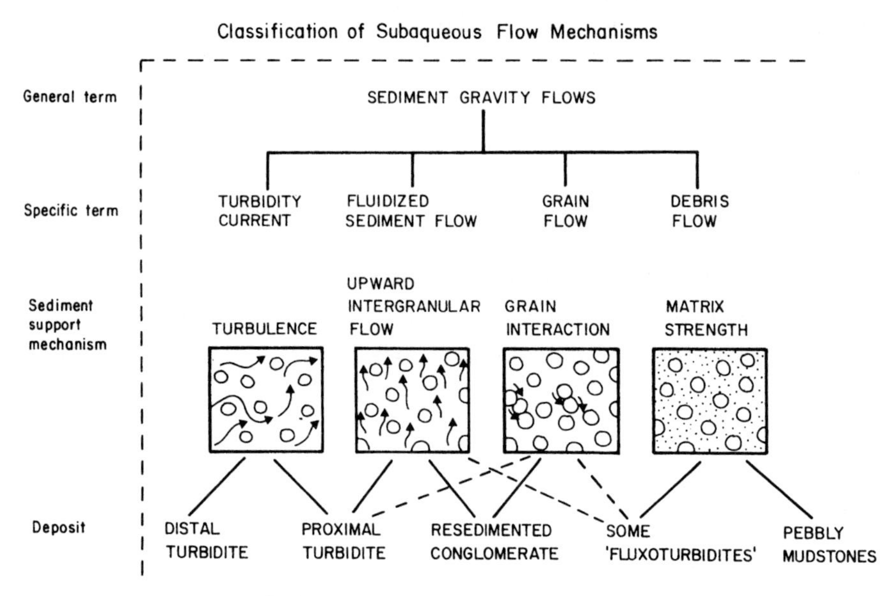


**Fig. 3.16** Classification of subaqueous sediment gravity flows (Middleton and Hampton, 1976). (*Reproduced by permission of John Wiley & Sons Inc.*)

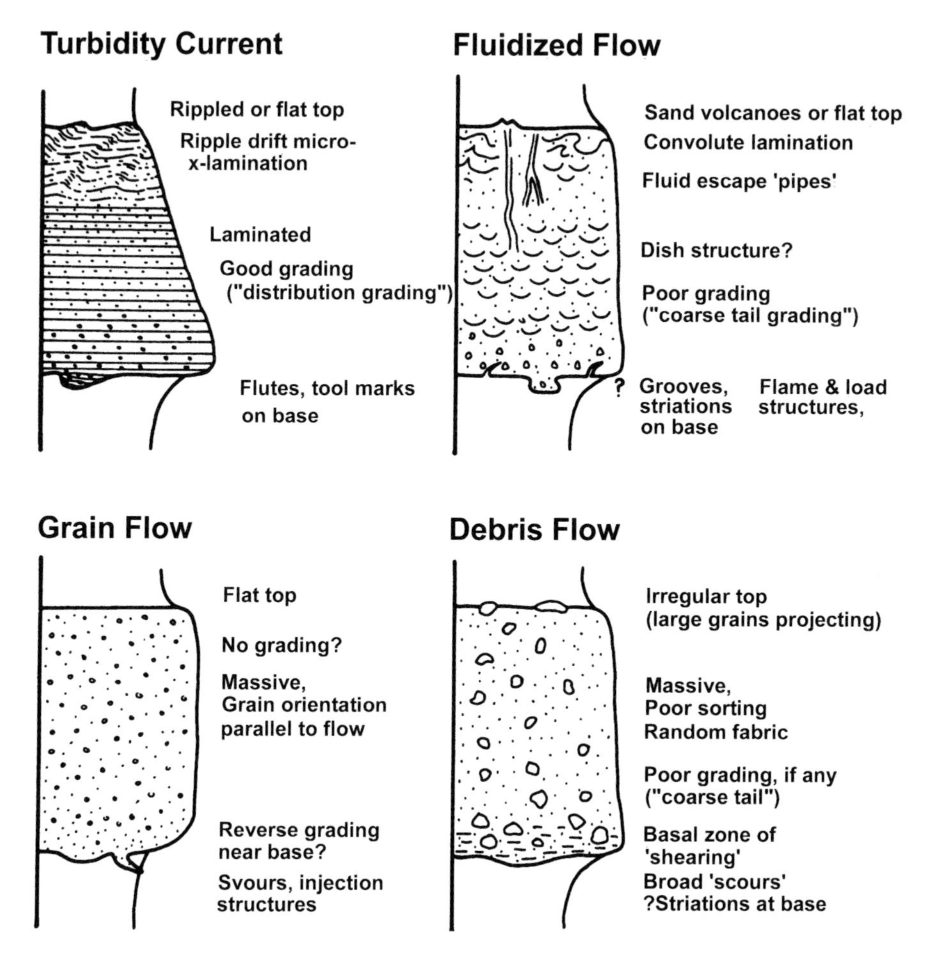


**Fig. 3.17** Sedimentary structures of gravity flow sediments (Middleton and Hampton, 1976). (*Reproduced by permission of John Wiley & Sons Inc.*)

are supported by upward fluid intergranular flows, *grain flow*, in which the sediment-supporting mechanism is dispersive pressure produced by interaction of grains, and *debris flow*, in which solid grains are supported by the clay matrix. The relationship between the four flow types and sediment types is shown in Figure 3.16 (Middleton and Hampton, 1976). The characteristic sedimentary structures produced by each type of sediment gravity flows were also summarized by Middleton and Hampton (1976) as illustrated in Figure 3.17.

In reference to grain flow, R. A. Bagnold (1956) found that dispersive pressure acts on each grain surface with a force in proportion to the square of grain diameter when cohesionless grains of various sizes are sheared, and that larger grains are pushed up to the top of the flow, which is the zone of least or zero dispersive pressure. As a consequence, larger grains are found in the upper parts of grain flows and debris flows (Okada, 1968b). The mechanism of the action of dispersive pressure is called the *Bagnold effect* (Sanders, 1963).

Further experimental studies of sediment gravity flows were carried out by Simpson (1987) and Mulder and Alexander (2001).

### 3.4.6 Turbidites and the Bouma sequence

Since the pioneering experimental work of Kuenen and Migliorini (1950), who demonstrated that turbidity currents produce graded beds, a large number of studies on turbidity current deposits of various geological ages were carried out by Kuenen and many other researchers. As already mentioned, Kuenen (1957) proposed the origin – indicating term *turbidite* for the deposits from turbidity currents.

The discovery that each turbidite bed may consist of five sequential parts, called *intervals*, distinguished by sedimentary structures (Fig. 3.18) was made by Arnold H. Bouma (Fig. 3.19), a professor of Louisiana State University, in his detailed sedimentological survey of the upper Eocene flysch sediments in the French Alps Maritimes, the Upper Cretaceous Gurnigel Flysch in Switzerland, the Upper Devonian of the Ardennes in Belgium, and the Devonian in West Germany (Bouma, 1962). It was his doctorate study as submitted to the University of Utrecht, the Netherlands. Walker (1965) suggested that the term *interval* be replaced by the term *division* in order to avoid confusion with time intervals. The five divisions recognized by Bouma, in ascending order, are as follows:

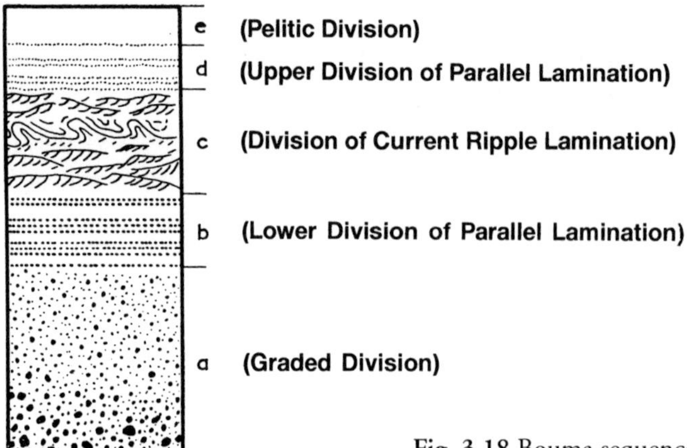


Fig. 3.18 Bouma sequence (Bouma, 1978).

**Fig. 3.19** Arnold H. Bouma (photo by H. Okada).

**Graded or Massive Division (Ta)**: This division shows graded bedding, and sometimes presents reverse grading at the lowest part of the division probably due to a traction carpet caused by grain collisions.

**Lower Division of Parallel Lamination (Tb)**: This is the division characterized by parallel lamination.

**Division of Current Ripple Lamination (Tc)**: This division is characterized by several sets of unidirectional foreset bedding; sometimes they are either replaced by climbing ripples or deformed into a convolute structure.

**Upper Division of Parallel Lamination (Td)**: This division is composed of fine sand and silt with parallel lamination.

**Pelitic Division (Te)**: The uppermost division consists of pelitic or clay material and shows no visible primary sedimentary structure. Indeed, Te may represent the normal condition of sedimentation in the basin.

These five divisions always indicate the vertical succession of a complete or ideal turbidite unit with a continuous relation between adjacent divisions. This complete sequence is called the Bouma sequence or the Bouma model (Bouma, 1978).

It would appear that the complete Bouma sequence is rather rare in occurrence. Depending upon the strength of the turbidity current and contained particle sizes, either the lower divisions of the Bouma sequence are predominant, and the upper divisions are eroded by succeeding turbidity currents, or they are missing. Hypothetical patterns of the lithologic changes of turbidites in a down-current direction are given in figure 3.20. It clearly illustrates that, as the velocity of turbidity currents decreases, not only the thickness of turbidites gradually decreases, but also the upper divisions of the Bouma sequence become predominant.

The concept of the Bouma sequence of turbidites has been applied to the palaeogeographic reconstruction of sedimentary basins. Based on the recognition that, when the Ta (or A) division of turbidites is absent, they start with the Tb (or B) division which represents a more distal deposition than those with Ta; when Ta and Tb are absent, Tc (or C) represents a still more distal position.

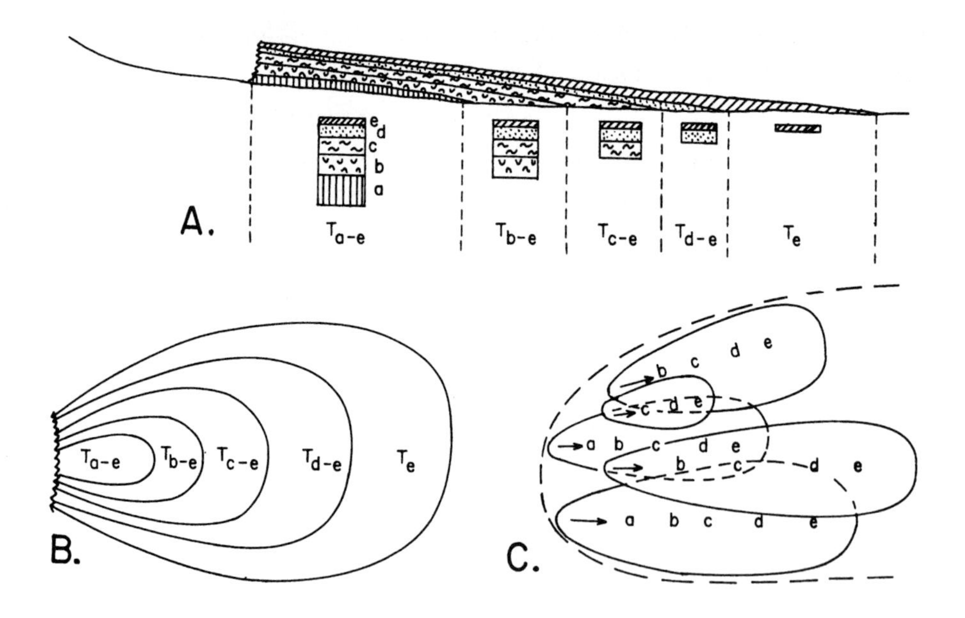


**Fig. 3.20** Hypothetical changes of the Bouma sequence in downcurrent direction (Bouma, 1978). A: Decrease in thickness and completeness of the sequence. B: Plan view of depositional cone with change of the sequence. C: Basin-filling by turbidites with the Ta to Te division at the bottom of a turbidite.

**Fig. 3.21** Roger G. Walker (photo by H. Okada).

Roger G. Walker (1965, 1967) (Fig. 3.21), a professor of McMaster University, proposed a quantitative index to show the relationship of proximal and distal depositional environments of turbidites by the proportion of beds beginning with divisions A, B, and C of the Bouma sequence (Fig. 3.22). This is the ABC Index or Proximality Index (P), which is expressed by:

$$P = A/(A + C) \times 100 \text{ or } P = A + B/2$$

where A, B, and C represent the percentage of beds starting with Ta, Tb, and Tc respectively.

Walker (1978) further classified the turbidite facies as a genetic scheme into:

(1) classical turbidite facies,

(2) massive sandstone facies,

(3) pebbly sandstone facies,

(4) grain-supported conglomerate facies, and

(5) pebbly mudstone facies.

The classical turbidite facies indicates an alternation of approximately equal amounts of sandstone beds with the Bouma sequence and mudstone beds. Mutti

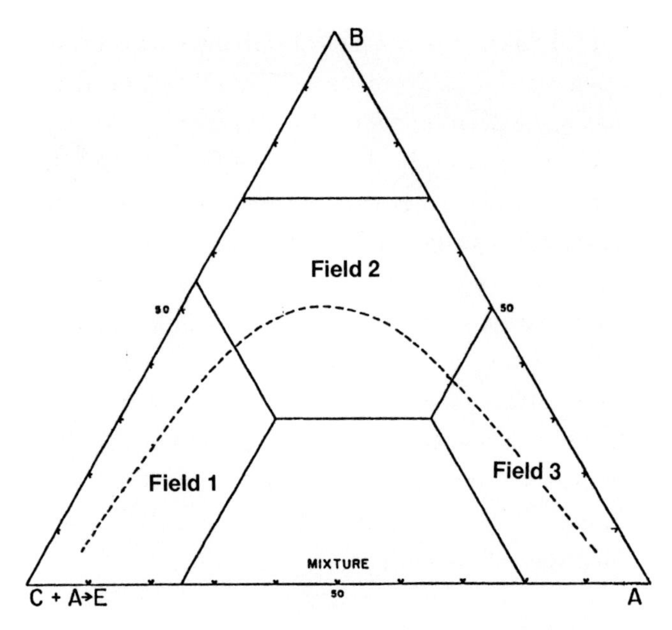


**Fig. 3.22** Proximality diagram of turbidites (modified from Walker, 1967). A, B and C are the Ta, Tb and Tc divisions of the Bouma sequence, respectively. Field 1 corresponds roughly with the lower flow regime of Simons *et al.* (1965), and Fields 2 and 3 with the upper flow regime, and 3 represents a higher regime than 2.

(1992), however, suggested the abandonment of the term 'classical turbidites' and instead the use of the term 'turbidites', which contain members of a broader range of genetically linked groups.

The recognition of the Bouma sequence in the turbidite facies helps to determine the areal pattern and spread of turbidity currents, permitting basin analyses of the deep-sea turbidite facies, a major theme of sediment studies in the 1960s. In addition, knowledge of the topography and sedimentary features of submarine canyons, through which turbidity currents flowed, has improved our understanding (Whitaker, 1974, 1976).

### 3.4.7 Geological significance of turbidites

The notion that turbidites characterize the basic sedimentary features of the flysch facies of orogenic belts was established in the 1950s and 1960s. Further developments in turbidite studies in the 1970s were inspired by investigations of the occurrence and origin of modern deep-sea turbidite deposits at the base of the continental slope and their application to ancient flysch facies. This implied that the main depositional environments of turbidites are equivalent to modern submarine fans. It was W. R. Normark (1970, 1978) who first proposed the submarine fan model or deep-sea fan model of turbidites.

Submarine fans are usually developed at the foot of the continental slope making fan-shaped rises or cones that can often be connected to submarine canyons. They are also encountered on the slope of island arcs. The cones thus contain terrigenous sediments that were transported by turbidity currents through submarine canyons. On an average fan sediments are about 0.5 to 3 km thick, but, in the largest instance, the Bengal Fan in the Bay of Bengal of the Indian Ocean is 1000 km wide, 3000 km long, and 12 km thick (Curray and Moore, 1971). For such fan-shaped morphology, Normark (1978) recognized three main subenvironments from the fan apex to the downslope bottom: the upper fan, mid fan, and lower fan (Fig. 3.23). Since the recognition that the main depositional sites of turbidites are in the submarine fan environment, a

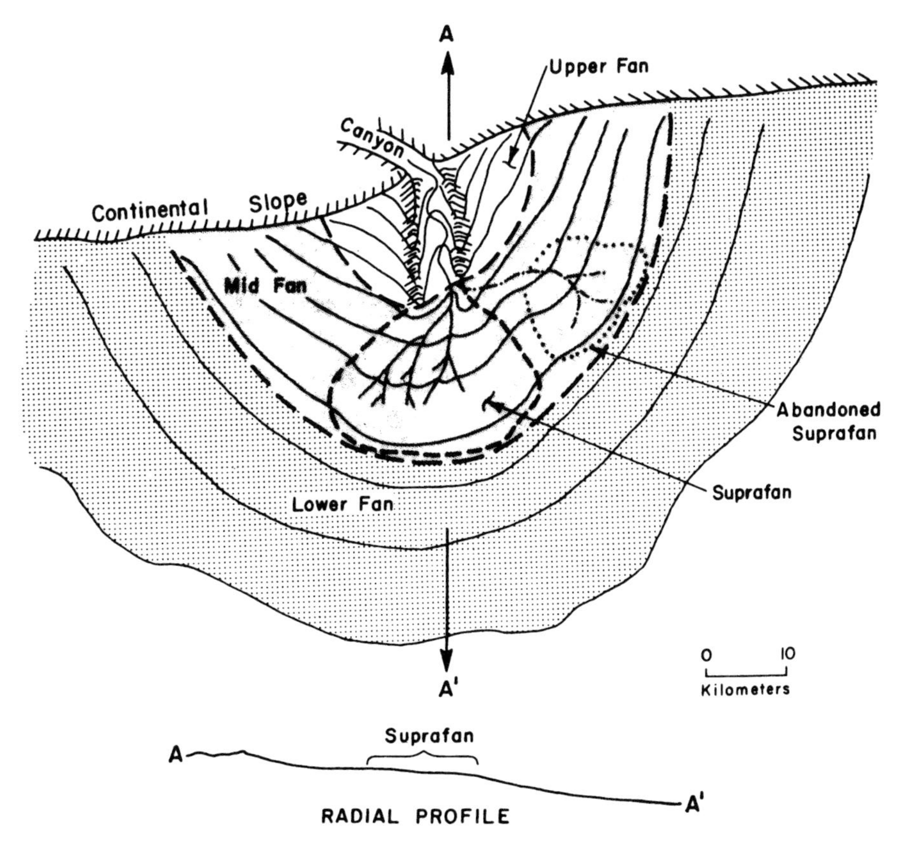


**Fig. 3.23** A submarine fan model (Normark, 1978). (*Reproduced by permission of the A.A.P.G.*).

number of studies of modern submarine fans were carried out in the 1980s and 1990s. A good example of a study of ancient fans was presented by Emiliano Mutti (Fig. 3.24), a professor of the University of Parma and Franco Ricci-Lucchi, a professor of the University of Bologna in Italy (Mutti, 1974; Mutti and Ricci-Lucchi, 1972). They recognized two major, if hypothetical, vertical facies sequences for progradational fan deposits: a thickening- and coarsening-upward facies sequence and a thinning- and fining-upward facies sequence. The former implies the deposition on the outer fan and mid fan, while the latter indicates the deposition in the channelled portion of suprafan lobes on the mid fan and deposition on the upper fan. The fan model by Mutti and Ricchi-Lucchi (1972) propelled the study of turbidites into a new epoch.

The geological significance of flysch turbidites was discussed and summarized by Harold G. Reading of Oxford University (Fig. 3.25) (Reading, 1972). The development of turbidite studies was reviewed by G. Kelling (1964) in Britain and by E. F. McBride (1964) in the United States as well as by Mutti (1992) and G. Shanmugam (2000, 2002) in a general review.

**Fig. 3.24** Emiliano Mutti (courtesy E. Mutti, 2001).

**Fig. 3.25** Harold G. Reading (photo by H. Okada).

### 3.4.8 Some problems in turbidite studies

The idea of reflected turbidity currents gave a new impetus to the continuing study of turbidites. Parea and Ricci-Lucchi (1975) found current ripples with directions at a 180° reversal from flute mark directions in the same bed in the Miocene Marnoso-Arenacea Formation of the northern Appenines which suggested up-fan climbing flows followed by a return to flows in previous directions. Kevin T. Pickering of the University of London and Richard N. Hiscott of St. John's University in Newfoundland, Canada (1985), also observed sedimentary structures showing a current direction at a 180° reversal to the main flow direction in turbidites in the Ordovician Cloridorme Formation of the Gaspé Peninsula, Quebec, Canada, and suggested that the reverse flow was due to reflection (deflection, even reversal) of turbidity currents from those which would have been expected (Fig. 3.26). They called such reflected turbidity currents *contained turbidity currents*. Although Pickering and Hiscott (1985) used two terms 'contained' and 'reflected', the

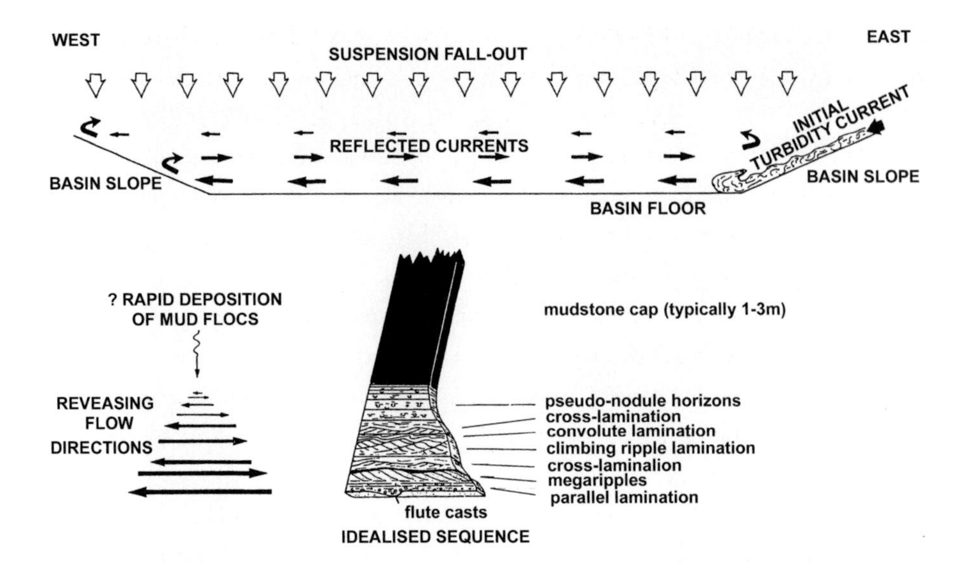


**Fig. 3.26** Deposition from a reflected turbidity current (Pickering and Hiscott, 1985). The upper figure indicates the process of reflection of a turbidity current on the basin-margin slope. The lower figure shows the turbidite sequence with changes of sedimentary structures corresponding to the initial flow and reflected flows. (*Reproduced by permission of International Association of Sedimentologists.*)

alternative term *reflected turbidity current* is more easily accepted here. Stimulated by these discoveries, the study of the reflection of turbidity currents has advanced (Pantin and Leeder, 1987; Edwards, 1993). It thus seems that the study of turbidity currents has entered a new stage of enquiry. Indeed, upslope turbidity currents have been reported on the Ceara Rise in front of the Amazon Cone, where the Rise stands as a barrier against turbidity currents from the Amazon submarine canyon (Damuth and Embley, 1979).

Clearly, more detailed studies of the behaviour of turbidity currents are required.

## 3.5 Seismic stratigraphy

The discipline of stratigraphy is mainly defined by lithostratigraphy, biostratigraphy and chronostratigraphy. Seismic stratigraphy is the study of stratigraphic and depositional facies based on seismic data. In order to describe depositional environments in relation to wave base, J. L. Rich (1951) introduced the terms *undaform, clinoform* and *fondaform,* as described in more detail in Chapter 5.3. It was mainly based on geophysical data for the investigation of subsurface geology. This may well be the discovery which, along with L. L. Sloss (1963) on major sediment packages on cratons, was an essential precursor to seismic stratigraphy.

Following the establishment of the geophysical technique of a continuous seismic reflection survey, it has become possible to observe the three-dimensional distribution of subsurface strata and their vertical sequence. This technique was first applied in submarine petroleum exploration. Now the stratigraphic method has evolved from correlation work on lithostratigraphy based on columnar data obtained from cores recovered at selected sites to the continuous trace of strata not only horizontally but also in the third dimension using seismic reflection profiles.

In order to construct a stratigraphy based on acoustic data, key reflector structures showing the successional relationship must be recognized on seismic reflection profiles. This is the fundamental requirement for establishing a stratigraphy based on the contact relationships between strata as seismic reflectors. It is possible to recognize subsurface stratigraphic units, to recognize and predict sedimentary environments through comparison between seismic stratigraphy and actual strata, and to analyse the relationship between seismic stratigraphic structures and sea-level changes. The science of the analysis of strata in such a way is called seismic stratigraphy. Examples of practical case studies of seismic stratigraphy were compiled by C. E. Payton (1977).

Seismic reflection profile data contain details of bedding surfaces, unconformity surfaces, and the continuity of lithofacies, bedding and bed thickness. The basic patterns of these features are shown by the continuity of the reflection and the boundary shapes of each reflection surface (Fig. 3.27). Well-stratified sediments show a continuous reflection with well-traceable parallel structures, whereas homogeneous massive sediments show chaotic or contorted reflections and/or no reflection. The basic styles of reflections in seismic stratigraphy are represented mainly by onlap, downlap, toplap and offlap (Fig. 3.27).

Onlap is the contact relation between the bottom profile of a basin and the horizontal basin-filling strata (Fig. 3.27A). The younger layers successively extend more widely than the distribution of the lower layers. Such a structure is characteristic of the lower surface of layers of basin-fill sediments. It is also important to accept that onlap reflects a relative rise of sea level.

Downlap is the style where the down-end of inclined strata contacts the horizontal surface of older sediments, showing a convergent relation between the progradational strata and the basement (Fig. 3.27B).

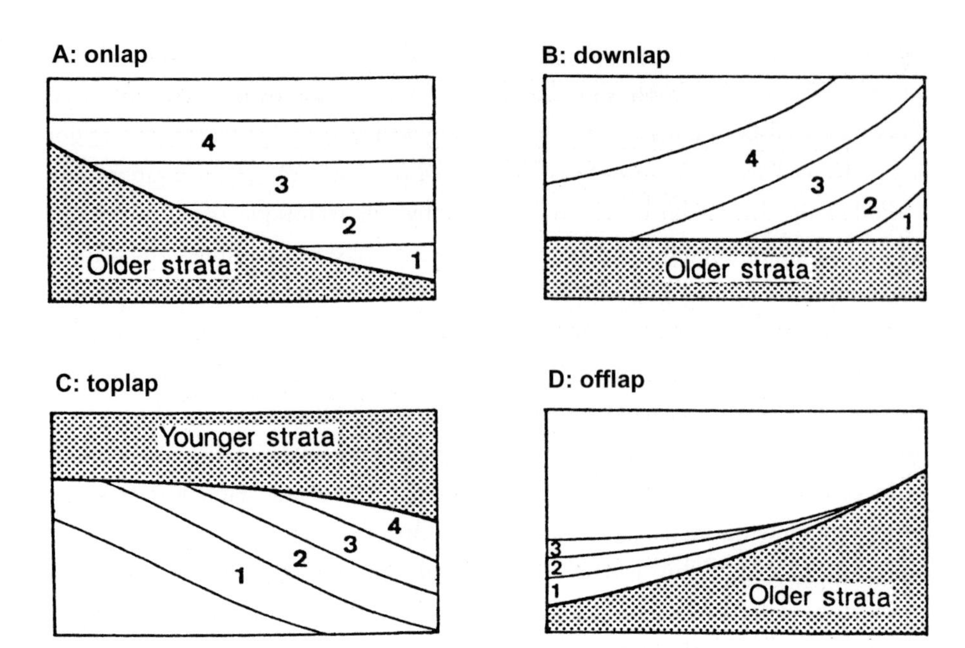


**Fig. 3.27** Basic patterns of seismic reflection boundaries 1–4 show the order of deposition. (*Reproduced by permission of the American Association of Petroleum Geologists.*)

Toplap is the style in which the uppermost part of inclined strata is suddenly cut and covered by horizontal strata (Fig. 3.27C). This is the structure of unconformity. Toplap follows minor erosion or non-deposition and major erosion with large relief giving an erosional truncation.

Offlap is a pattern of reflectors where the upper end of inclined strata migrates towards the central part of a basin and younger beds are, therefore, progressively further away from the basin margin (Fig. 3.27D). This phenomenon suggests a regression with lowering sea-level.

When certain groups of sediments are analysed on the basis of these reflection structures, such successional units of sediments as indicated in Fig. 3.28a can be discriminated. These successional units are either bounded on the upper and lower surfaces by unconformities in the areas of shallow sea and on land, or separated from upper and lower units by correlative conformable boundaries in the deep-sea areas. These successions within units are related to each other in origin. The geometric extent of these units through time is shown in Fig. 3.28b.

Worthy of note in Fig. 3.28a is that the lower part of a unit sequence shows an erosional clino-unconformity; when traced laterally, it shows a conformable relationship on the slope area between the shallow-sea and deep-sea areas; and

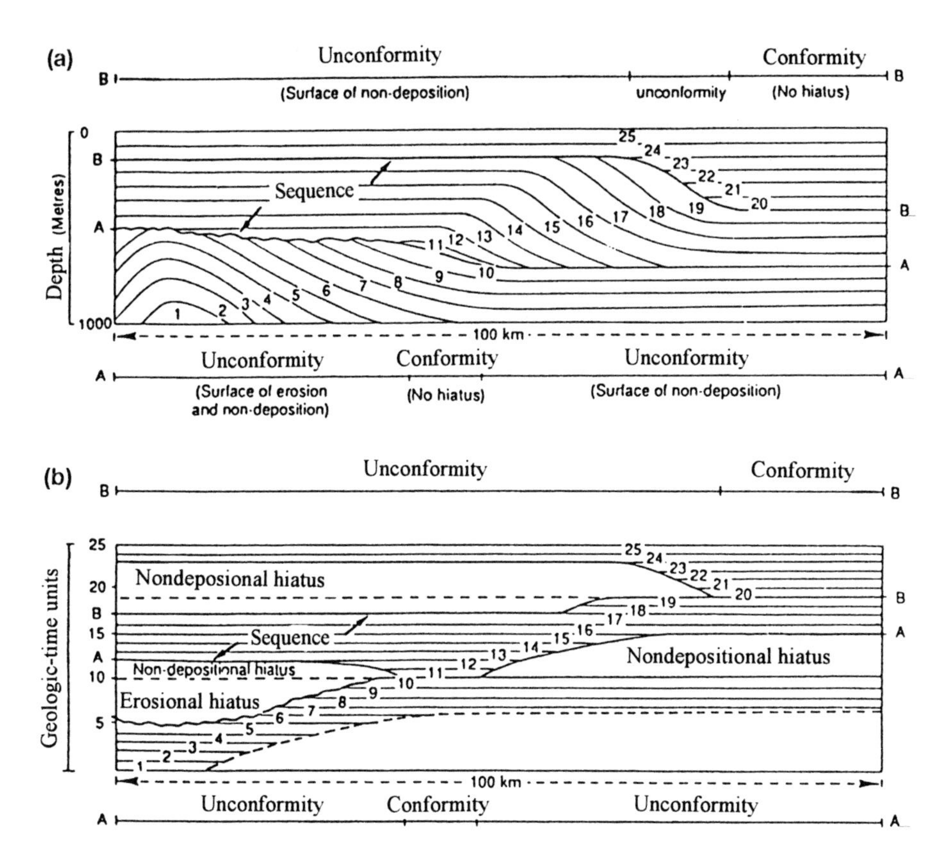


**Fig. 3.28** Basic concept of depositional sequences (Mitchum *et al.*, 1977). (a) Generalised stratigraphic section. (b) Generalized chronostratigraphic section of the same section as in (a). A – A, B – B: sequence boundary. (*Reproduced by permission of the A.A.P.G.*)

further to the right it is changed into a nonconformity due to non-deposition in the deep-sea area. This illustrates that the time interval at the unconformity position is also recorded in the conformable sequence.

In order to recognize sequence units, two kinds of unconformity (type 1 and type 2) are distinguished (Fig. 3.29). A time when relative sea level falls below the shelf edge is clearly indicated in continental margin sediments. If relative sea level falls below the shelf edge, the type 1 unconformity is produced, and when sea-level fall is insufficient, the type 2 unconformity is formed.

The type 1 unconformity is produced by the erosion of the land and its offshore continental shelf margin. In other words, this type of unconformity indicates erosion in both on-land and submarine environments. This is caused by a fall of sea level that exceeds the rate of subsidence of a sedimentary basin. This type

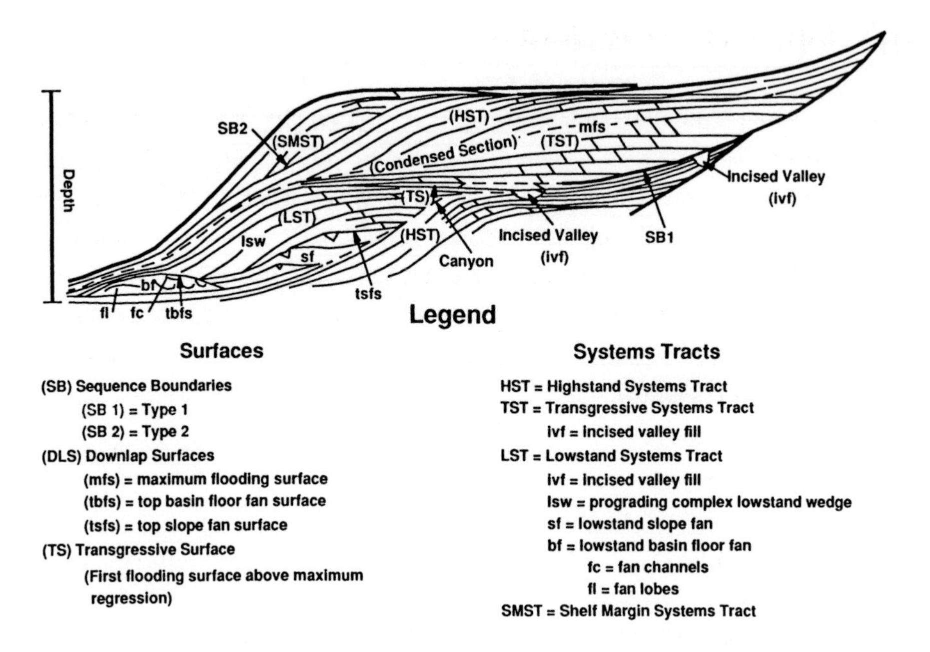


**Fig. 3.29** Sequence stratigraphy depositional model showing sequences and facies systems tracts (Armentrout, 1990).

of stratigraphic boundary is called the type 1 sequence boundary in sequence stratigraphy.

Conversely, the type 2 unconformity shows small erosion rates in the on-land area and no submarine erosion. This is caused when sea level falls more slowly than the basin-floor subsidence. This type of stratigraphic boundary is called the type 2 sequence boundary in sequence stratigraphy.

A sedimentary unit bounded by types 1 and 2 unconformities is built up under continuous sedimentary processes and is defined as a sequence. Thus, the basic approach to seismic stratigraphy is to identify the seismic sedimentary sequence, which is further refined by the analysis of seismic lithology, the reconstruction of sedimentary environments, and the recognition of sea-level changes. This methodology constitutes the basis of sequence stratigraphy (Vail *et al.*, 1977a).

The advances in allied subject areas and their influence on the main theme of sedimentation also need some attention. One of the most influential works was by Leopold *et al.* (1964) which revolutionized geomorphology by its remarkable advances in the application of quantitative and analytical approaches. It emphasized geomorphic process in erosional debris and running water acting to shape the land surface. It has had a profound effect on sedimentology.

## 3.6 Sequence stratigraphy

Sequence stratigraphy constructed on seismic stratigraphy is a way of discriminating sedimentary successions or packages with similar sedimentary features based on unconformities, and of analysing sedimentary environments of the whole sedimentary succession. As to the importance of such sedimentary sequences bounded on both the upper and lower surfaces by unconformities, L. L. Sloss, a professor of Northwestern University, first recognized six sequences separated by unconformities in the Precambrian to Palaeocene sediments exposed in the interior stable area of the North American Craton (Sloss, 1963), and K. H. Chang, a professor of Kyungpook National University, Taegu, Korea, has recognized four sequences bounded by unconformities in the Palaeozoic to Mesozoic basins of the Korean Peninsula. Chang (1973, 1976) proposed a new concept of 'Synthem' for a sedimentary sequence bounded on both the upper and lower surfaces by unconformities, giving a new aspect to stratigraphy. It was, however, Peter Vail (Fig. 3.30), the leader of the Exxon Research Group and later a professor of Rice University in Houston, Texas, who not only contributed to the establishment of seismic stratigraphy, but also proposed a new concept for

**Fig. 3.30** Peter Vail (photo by H. Okada).

stratigraphy as a systematically organised theory of sequence stratigraphy. He and his colleagues started their research of sequence stratigraphy in 1987.

The idea of sequence stratigraphy was at first very theoretical in that it was postulated to be controlled by eustatic sea-level change (Mitchum *et al.*, 1977). It was successfully developed, however, through a detailed combination of seismic stratigraphy and lithology, sedimentary environments and dating to show that depositional environments of sedimentary sequences can be correlated with relative sea-level changes (Vail, 1987). It is reasonable to argue that sequence stratigraphy has brought about a revolution in stratigraphy since the 1980s. Behind this trend, there seems to have been the expansion of thinking on the relationship between sedimentation and sea-level changes and the establishment of chronostratigraphy. Andrew D. Miall, a professor of the University of Toronto, and Charlene E. Miall, a professor of McMaster University, considered that sequence stratigraphy is indeed a new paradigm in geological thought (Miall and Miall, 2001), comparable to the scientific revolution proposed by Thomas Kuhn (1962), a professor of the history of science at Princeton University.

### 3.6.1 Concept of sequence stratigraphy

Sedimentary successions consist of sedimentary systems with three dimensional lithologic features called sequences. There are two significant factors controlling sequences, namely *accommodation space* and *sediment supply*.

#### *Accommodation space*

Sediments are accumulated in a space where they can be deposited and preserved. This space is referred to as a sedimentary basin. The sediments that filled a basin are called the basin-fill succession. At the time of sedimentation in a sedimentary basin, any spaces for sedimentation are called the accommodation space (Jervey, 1988). The accommodation space in shallow-sea environments is generally created either by a rise of sea level or subsidence of the sea floor. In such cases, the upper limit of the accommodation space is controlled by sea level. It is obvious, however, that sea level itself will not directly control the accommodation space in the deep-sea environment.

The accommodation space in the on-land area, hence above sea level, is sited in river channels and flood plains and may be controlled by the local gradient of channels. The accommodation space in lacustrine environments is controlled by both the lake-water level and the lake floor.

### *Relation between rates of accommodation space formation and those of sediment supply*

There are some distinct relationships between the rate of formation of the accommodation space and the rate of sediment supply. If the rate of formation of the accommodation space and the rate of sediment supply are in balance with each other, sediments will build up vertically; namely, aggradation will take place. In this case, the coastline does not shift, but is fixed. If the rate of sediment supply is larger than the rate of formation of the accommodation space, sediments will grow basinwards and upwards giving progradation. In the seacoast area, the seashore will move seawards and the coastline will also move seawards. In this case, the vertical lithofacies of seashore sediments shows the deposition of foreshore sediments, showing also a trend of shallowing-upward sedimentation. This is the expression of the regression. If the regression is caused by a fall of sea level, this phenomenon is defined as forced regression (Posamentier *et al.*, 1990, 1992). If the rate of formation of the accommodation space exceeds the rate of sediment supply, seashore sedimentation will be out of phase with the rise of sea level, and seawater will flood the shore area. The coastline will shift landwards; this is retrogradation and a transgression will take place. The lithology in this case shows a deepening-upward sedimentation.

### *Sequence and systems tracts*

Stratigraphic features derived from the balance between the sediment supply and the accommodation space constitute the basic pattern of basin-fill sediments. A sedimentary pattern during a particular period of sedimentation in a basin is referred to as the depositional sequence. The main feature of this pattern is that a unit sequence is bounded at both the upper and lower surfaces by an unconformity that changes oceanwards into conformity. This pattern indicates sedimentation during a time between two low sea level periods. The sedimentary surface in the period of sea level fall defines the sequence boundary.

The time between the two low sea level periods is a period of sea level rise. The sediments during this period show evidence of transgression, aggradation, and progradation, in accord with changes of sea level, each giving different sedimentation patterns. The sedimentary units from similar sedimentary processes are referred to as the systems tract.

At the time of high sea level, shallow sea areas are covered by deeper water facies and landward shifts of the lithology take place. The lithology with such features is referred to as a transgressive systems tract. As the sea level gets much higher, surface sediments will be removed by wave erosion at the shore line as well as by subaerial erosion. The erosion surface produced by such transgressions is called the transgressive erosion surface or ravinement surface (Swift, 1968).

When the speed of sea level rise decelerates and the accommodation space is balanced by the rates of sediment supply, aggradation will predominate. When the sea level reaches its highest position and no more accommodation space is produced, progradation will take place. Sedimentary profiles at the time of high sea levels show either aggradation or progradation, as indicated above, constructing a highstand systems tract.

When the sea level drops, the landward continental shelf is exposed and the sequence boundary is created by erosion. This erosion process will continue until the sea level drops to the lowest position. When the sea level falls to the margin of the continental shelf, the whole shelf area is exposed and sedimentation proceeds both on the shelf (fluviatile, aeolian, lacustrine) and on the slope of its outer margin. This is called the lowstand systems tract.

The thickness of a depositional sequence depends upon the productivity of depositional space in a basin. As sea-level changes are controlled by both tectonic and thermo-tectonic movements of basements in addition to eustatic controls, depositional sequences range in thickness from a few metres to thousands of metres.

The succession of progradation in a highstand systems tract can be grouped into some minor series showing a coarsening-upward shallowing sequence. Such minor sequences with a shallowing-upward trend are called parasequences. A parasequence is, in general, composed of deposits of shallow-marine environments a few metres to tens of metres thick, with offshore muddy sediments at the basal and lower parts and with shallow-sea sandy sediments at the upper (Fig. 3.31).

A depositional sequence consists of parasequences, which can be considered to reflect changes of river channels. These successions with parasequences were previously referred to as the concept of the 'sedimentary cycle'.

To sum up, sequence stratigraphy studies the genesis of sequences by means of correlation with sea-level changes. In this respect, it should be pointed out that on-land sediments themselves cannot be expected to correlate with sea-level changes. However, the following hints may help. When sea level falls, the down-cutting of rivers is accelerated to develop channel sediments, whereas when sea level rises, floods frequently occur and develop floodplain sediments. Furthermore, at the time of sea-level fall, groundwater levels also fall tending to produce soils that exhibit oxidation.

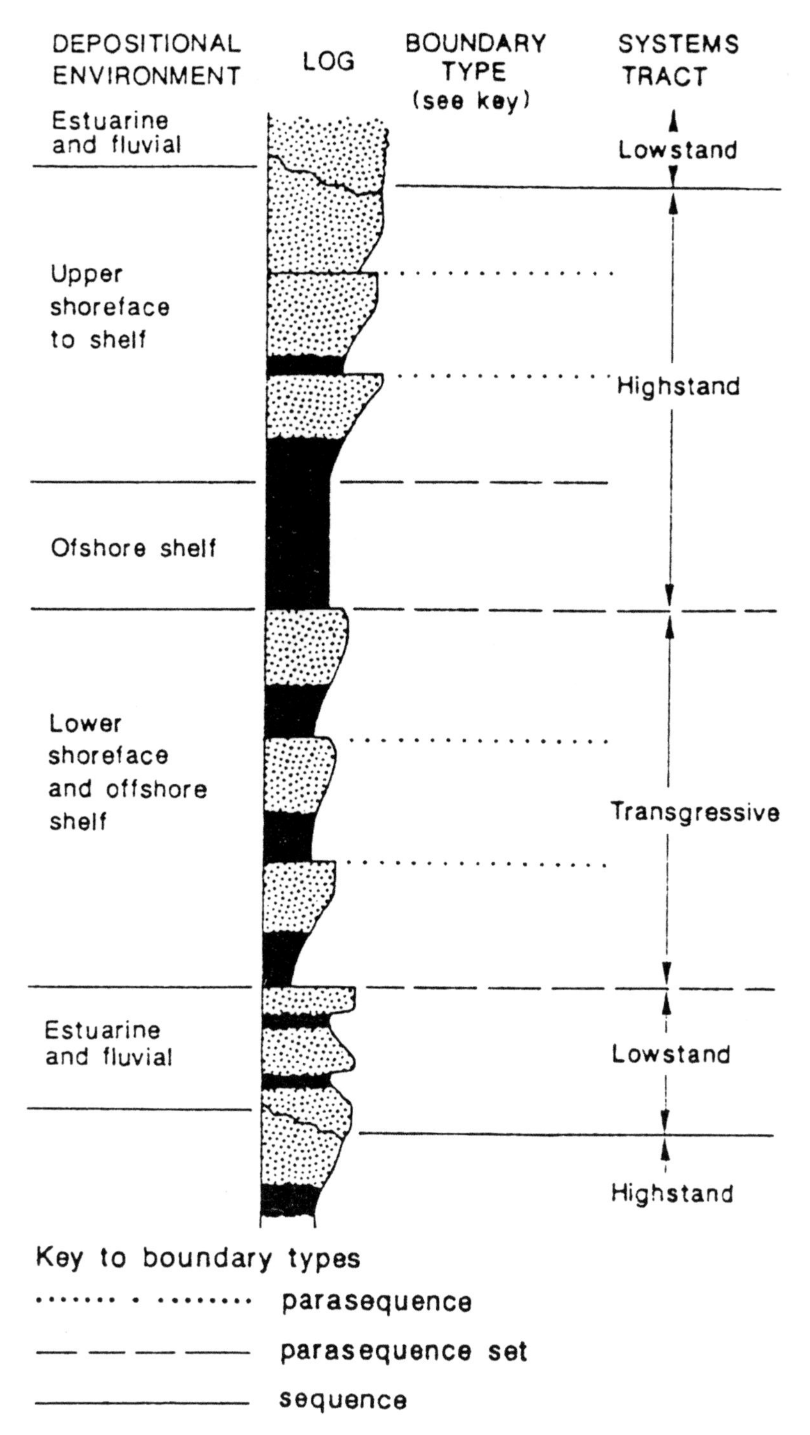


**Fig. 3.31** Relation between sequence and parasequence (modified from Wilson, 1991). (*Reproduced by permission of the Geological Society of London.*)

### 3.6.2 Cycles of sea-level changes and sequences

In this section it is intended to describe the relationship between the rise and fall of sea level and variations in sequences in the profile from on-land fluvial environments to deep-sea environments.

*Falling sea-level stage (regressive stage)*

Subaerially, river systems establish a new dynamic equilibrium according to the sea levels of the falling stage. The areas covered with sea waters at the high stand become exposed and the down-cutting of rivers starts. Shallow-sea areas migrate offshore, and get narrower as the sea level falls to the continental shelf margin. In the tropical shelf areas, fluvial clastic sediments with coral-reef limestones are developed, in which karst topography is formed. Increase in erosion of on-land areas and decrease in accommodation space produce large amounts of clastic sediments which are transported beyond the edge of the continental margin and deposited on the deep-sea floor. This transportation is made mainly by turbidity currents which develop submarine fan turbidites with lobe deposits. Thus, the sequence boundary in the deep-sea environments is indicated by a progradational surface of coarse-grained turbidites.

*Rising sea-level stage (transgressive stage)*

As the sea level gradually rises from the lowest level, deposition takes place in down-cut rivers, showing aggradation. At the same time, areas of down-cut river mouths are drowned and become sites of marine deposition. The existence of down-cut rivers filled with fluvial and river-mouth sediments on the shelf is an indicator of a sequence boundary. Landwards from the outer shelf edge, shelf environments proceed with the aggradation of relatively fine-grained sediments.

In the fluvial sediments system, down-cut rivers are filled by aggradation and flood sediments are widespread in the flood plain. The seashore areas are covered by sea-water and marine sedimentary environments are expanded landwards. Thus sediments exhibiting retrogradation are characterized by coarse-grained facies in the coastal environments and do not reach the middle to marginal parts of the shelf. In this way, the uppermost part of a transgressive systems tract reduces sediment supply, producing a condensed facies.

*Highstand systems stage*

During the period of rising sea level, fluvial sedimentary sequences continue to aggrade. However, when the sea level gets to the top-stand position, vertical aggradations decline. The production rates of accommodation space in shallow-sea environments are equivalent to the rates of sediment supply, or are overcome by the latter. This results in the coastline shifting seawards. Therefore, as sedimentation proceeds onto the shelf, shallow-marine sediments become more predominant, terrigenous clastics spread over the shelf margin and then extend to the deep-sea environment.

### 3.6.3 Sequence stratigraphy and sea-level change

As described above, it is clear that the formation of sedimentary sequences is closely related to sea-level change. In general, the rates of rise and fall in sea level control the rates of shifting of onlap in the seashore areas. It is possible to recognize relative changes of sea level by identifying transgressive and regressive facies in lithologic columnar sections. If a sequence boundary is dated successfully, it is further possible to identify the time corresponding to sea-level change. In recent years, it has become possible to enhance the accuracy of dating sea-level changes by applying detailed biostratigraphic zonations based on large numbers of different groups of taxa, magnetostratigraphy, and chemostratigraphy using oxygen and carbon isotopes and other elemental compositions of deposits in addition to radiometric dating methods. The well-known Vail-Haq curve (Fig. 3.32) was prepared on the basis of such principles (Vail *et al.*, 1977b; Haq *et al.*, 1987).

It is interesting to note that the Vail-Haq curve is subdivided into three major types of sea level change: a first-order cycle (Fig. 3.32a), a second-order cycle (Fig. 3.32b), and a third-order cycle (Fig. 3.32c). The patterns of each cycle overlap each other on the first-order curve. The cyclicity of the major cycles is subdivided into six orders when natural phenomena are analysed in detail. The time length of each cycle and its origin are summarized in Table 3.1.

This table suggests that long-period cycles reflect changes in the mass volume of the mid-oceanic ridges and changes in velocities of ocean-floor spreading while short-period cycles are controlled by the development and decay of continental ice-sheets: both are important phenomena on a global scale. The following processes are important as the causes of sea-level change:

**Local tectonic movements**: The uplift and subsidence of the basement in the continental margin environments due to local tectonic movements, such as earthquakes, tend to cause vertical changes of sea level initiated by tsunamis. The sea-level change in such cases is a local phenomenon in time and space.

**Table 3.1** Orders of eustatic sea-level changes

| | Cyclicity | Possible causes |
|---|---|---|
| First order | 200–400 Ma | Tectonic-controlled sea-level changes: Formation and breakup of supercontinents (e.g. Pangaea) |
| Second order | 10–100 Ma | Tectonic-controlled sea-level changes: Volume changes in oceanic ridges |
| Third order | 1–10 Ma | Local tectonic-controlled sea-level changes |
| Fourth order | 0.1–0.5 Ma | Glacial-controlled sea-level changes: Late Carboniferous – Early Permian (340 – 270 Ma) Late Cenozoic (35 Ma) |
| Fifth order | 0.01–0.1 Ma | Glacial-controlled sea-level changes |
| | <0.1 Ma | Milankovich (1941) cycles |
| | 92,000 a | Eccentricity cycles |
| | 40,000 a | Obliquity cycles |
| | 21,000 a | Precession cycles |
| Sixth order | <0.01 Ma | |
| | 22 a | Sunspot cycles |
| | 11 a | Sunspot cycles |

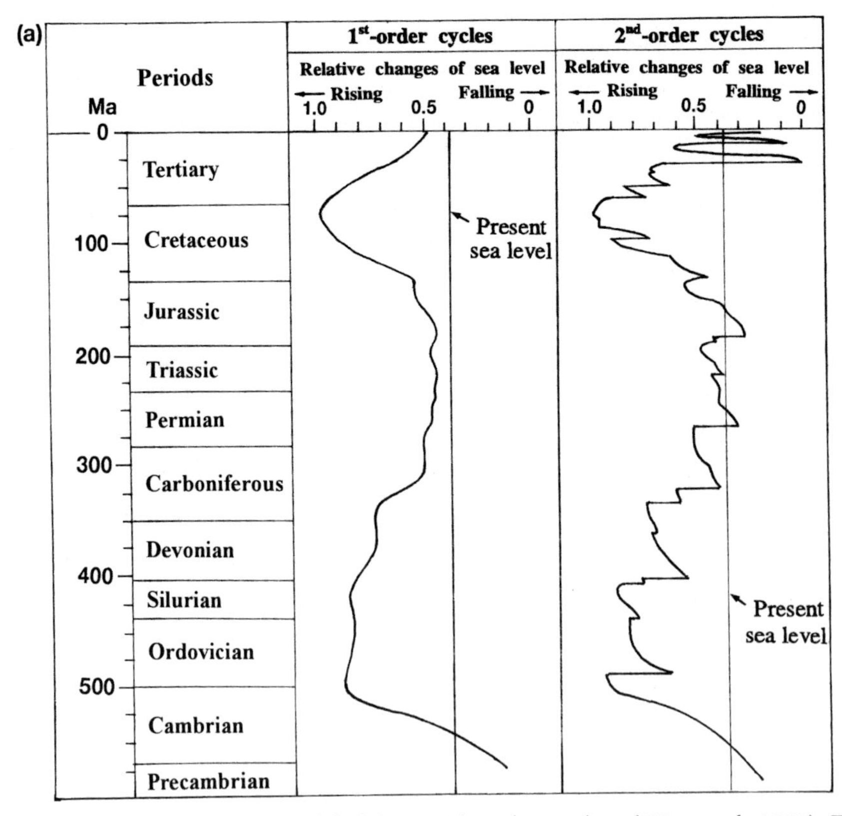


Fig. 3.32a Vail-Haq curves (modified from Vail *et al.*, 1977b and Haq *et al.*, 1987). First and second order cycles during the Phanerozoic. (*Reproduced by permission of the A.A.P.G.*)

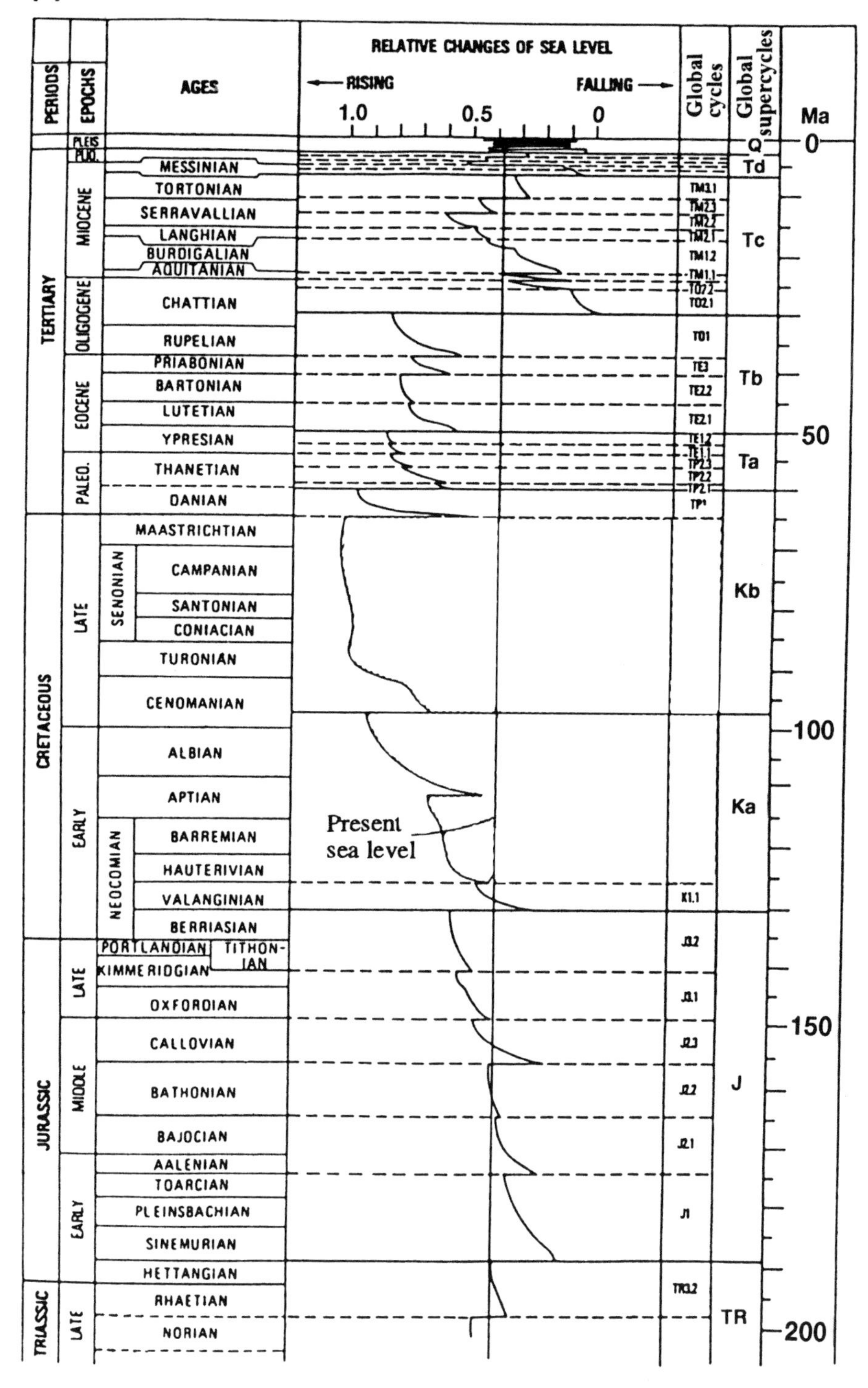


Fig. 3.32b Relative changes of sea level during the Jurassic–Tertiary.

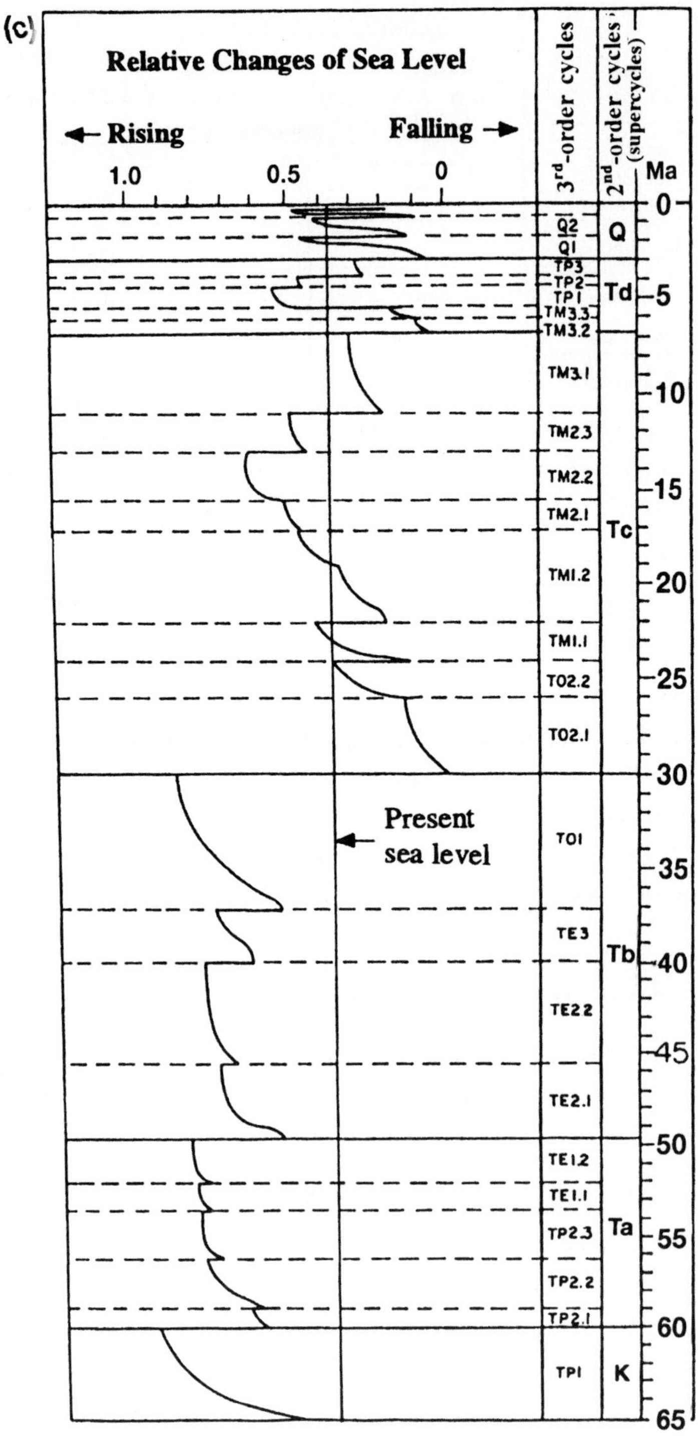


Fig. 3.32c Relative changes of sea level during the Cenozoic.

**Glacial control**: The most direct cause for the sea-level change is considered to be due to the development of glaciers and ice-sheets and their melt in high latitude regions. Case studies indicate that the freezing and melting of ice at both the Poles can cause a global rise and fall of sea level in the order of 100 m. According to Plint *et al.* (1992), if the whole ice mass of the Antarctic should melt, the global sea level rise will be about 150 m. The development and melting of polar ice directly control sea level, and such changes can be quite rapid. Sea-level changes in the Pleistocene glacial period range from some 10 thousands to 100 thousands years in duration.

**Thermal tectonics**: The rifting of the continental crust and collision of plates are associated with thermal changes in the earth interior. Such large-scale tectonic movements cause global sea-level changes.

**Wind force**: Strong wind drifting due to typhoons and other storms may cause local and short-lived changes of sea level as will changes in atmospheric pressure.

**Evaporation**: Lakes and bays separated from the open sea where evaporation exceeds the influx of land water and rain-water, cause sea-level to fall due to evaporation. A particulary extreme example of this process was the late Tertiary evaporation of the Mediterranean Sea (Hsü, 1983). Such was its scale that it must also have led to raised sea-levels in the global ocean due to increased precipitation outside the Mediterranean area.

### 3.6.4 Some problems in the application of sequence stratigraphy

Sequence stratigraphy is quite effectively applied to the analysis of sedimentary environments on the basis of relative sea-level changes, but it should be applied carefully.

#### *Checking the sequence stratigraphic models*

As the concept of sequence stratigraphy is to identify unconformities and their correlative conformities in the system of established chronostratigraphy, it is somewhat difficult to apply sequence stratigraphy effectively to deep-sea deposits in which terrigenous clastics are rare and the recognition of unconformities is very difficult. The same applies to continental deposits in which precise dating is difficult even if unconformities are easily identified. It should be borne in mind that, because the type model of sequence stratigraphy was established in the shelf to continental slope deposits of the western Atlantic where clastic sediments have been supplied almost continuously since the rifting of the Gondwana continent, revised sequence stratigraphic models should be constructed to be applicable to other regions with the proper identification of appropriate regional sea-level changes.

#### *Where the tectonic activity is intense*

Local sea-level changes should be taken into account in the application of sequence stratigraphy when uplift and subsidence due to tectonic movements are more significant than the influence of global sea-level change or eustacy. For example, in an area of active subsidence of the crust, sediments may generally show the features of the highstand systems tract, whereas in an active tectonic area, where uplifting of the crust exceeds the rise of sea level, active sedimentation continues, even if a high sea level is maintained. It is of interest to note in this respect that mountains, as high as 4000 m, at the eastern margin of the Asian continent supplied large amounts of clastics to the present Japanese Islands area in the Cenomanian to Campanian time of the Late Cretaceous (Okada, 2000), a time of the highest global sea level according to Vail *et al.* (1977b).

#### *In the case of non-marine deposits*

It is generally difficult to establish sequence stratigraphy in non-marine deposits unless there are intercalations of marine layers within them. This is because non-marine deposits themselves have rare index fossils effective for the accurate determination of times of sedimentation. If, however, such significant data are available, sequence stratigraphy is applicable to non-marine deposits through correlation with contemporaneous marine deposits. If tephra deposits exist, sequence stratigraphy may be more successfully applied.

The relationships between coal deposition and the sequence stratigraphic framework have been discussed by many workers (Aitken, 1994; Flint *et al.*, 1995; Holz *et al.*, 2002). Useful suggestions for resolving problems and difficulties in establishing sequence stratigraphy in non-marine deposits are summarized by Takano (2001).

#### *Globalization of stratigraphy*

Sequence stratigraphy, which lays stress on correlation with global sea-level changes, may be nothing less than the extension or globalization of local stratigraphy. That is to say, sequence stratigraphy is intimately linked to global stratigraphy (Wilson, 1998). How accurately sequence boundaries recognized in any particular area can be correlated to global sea level remains an important topic to be resolved in the future.

# 4

# Development of sedimentary petrology

## 4.1 Microscopic sedimentary petrography

After James Hutton, Charles Lyell, Georges Cuvier, and William Smith the stratigraphic studies of sediments developed rapidly, but almost no interest was paid to the mineralogical and petrological compositions of sediments. It was Henry Clifton Sorby (Fig. 3.2) who opened a new path in geological studies in this field.

Around 1848 Sorby learned from William Crawford Williamson, a physician in Manchester, a method of preparing thin, optically transparent, sections of bones and teeth. Sorby soon became skilful in this practice and applied the method to hard rocks. Sorby's first work using this method was in his study of the geology of the Malvern Hills, published in 1849 (Sorby, 1849): he was twenty-three years old. This was the first paper to apply thin section microscopy to geology (Higham, 1963), and the same methods of making thin sections of hard rocks are still used today. It thus became possible to observe thin sections of rocks using transmitted light through the microscope (Judd, 1908). It is said that Sorby had a great enthusiasm for the physical sciences, particularly in the field of optics and the microscope during his school age. This was a solid foundation for Sorby's future career (Higham, 1963). It is of interest to point out that William Nicol had already invented the polarizer or *Nicol's prism* for polarized microscopy in 1828. Sorby enjoyed the observation of minerals with varieties of optical properties under the polarizing microscope. After his paper on the Malvern Hills (Sorby, 1849), his study of sedimentary rocks using microscopic observations began with the Lower Calcareous Grits of the Yorkshire coast, and his paper was published in 1851 (Sorby, 1851). Thus, the field of microscopic sedimentary petrography started (Sorby, 1877).

In July 1861, Sorby attended a conference in Germany. On this tour he met a young student of geology, Ferdinand Zirkel (1838–1912), later a professor of Kiel University, and Sorby explained how to prepare thin sections of rocks and the progress he had made in microscopic petrography. This meeting brought about a dramatic development in microscopic petrography, mainly of igneous rocks, in Germany and other European countries, by Zirkel and by Karl Harry Ferdinand Rosenbusch (1836–1914), a professor of Heidelberg University. They together with Sorby, established microscopic petrography in the history of geology.

## 4.2 Sedimentary petrology

### 4.2.1 Sedimentary petrography

The microscopic sedimentary petrography of sedimentary rocks, which was inaugurated by Sorby, came into flower and full bloom in Britain. The first study in this new field, which started the epoch of sedimentary petrography, was the petrographic description of carbonate rocks by Sorby himself (Sorby, 1851, 1879) and also of non-carbonate rocks (Sorby, 1880).

Sedimentary petrography systematically describes the mineralogical composition of sediments and sedimentary rocks and the lithological features of rock specimens. Any review of this new development in this discipline of earth sciences must start in Britain. Fundamental petrological studies of sedimentary rocks in Britain are represented by Wm. Mackie and Albert Gilligan. Mackie (1896) divided the quartz in sands into four groups and showed possible source rocks for them. In 1919 Gilligan (1920) also followed Mackie's method of study and reached similar conclusions with regard to the Millstone Grit. These methods are still important at present.

The study of heavy minerals in sedimentary rocks pioneered by T. G. Bonney (1900) and developed by H. H. Thomas (1902) is a striking example of the application of microscopy to geology. Their new method of heavy mineral studies presented the opportunity to determine the composition of the source rocks (provenance) of clastic sediments, to reconstruct the palaeogeography and to correlate strata. It anticipated the future significance of the study of heavy minerals. Heavy minerals are defined as minerals with a specific gravity more than those of quartz (2.65) and feldspars (2.54–2.76), both of which are the major constituents of clastic sediments; generally the specific gravity of heavy minerals is greater than 2.8.

Later eminent studies of heavy minerals were carried out by I. S. Double (1924) with detailed descriptions of the mineral composition of Late Tertiary sediments in eastern England and by A. W. Groves (1931) on the relationship between the erosion of the Dartmoor granite of southwestern England and transport paths of the derived clastics. Textbooks, regarded as splendid contributions to the golden age of microscopic sedimentary petrography in Britain, are represented by F. H. Hatch and R. H. Rastall (1913; the first English textbook of sedimentary petrography, with revised editions being repeatedly published, the last by Greensmith, 1976), H. B. Milner (1922, 1929), and P. G. H. Boswell (1933). In addition, a subsequent textbook by Milner (1962) is an excellent compilation of sedimentary petrography which details the features of rock-forming minerals in their naturally occurring shape with a wealth of microscopic sketches; it still remains a splendid guide to the study of heavy minerals.

In France, a superb monograph of microscopic descriptions of French sedimentary rocks was compiled by Lucien Cayeux (1864–1944) (Fig. 4.1), professor of Collège de France (Cayeux, 1929a, b). There is also a masterpiece of petrographic descriptions of Swedish sedimentary rocks by A. Hadding (1929) and good contributions to the petrography of German sedimentary rocks by C.W. Correns (1939, 1949) and H. Müller (1933).

**Fig. 4.1** Lucien Cayeux (*reproduced by permission of Société Géologique de France*).

The United States of America also made major contributions to sedimentary petrography. M. I. Goldman (1915) took the initiative in the study of sedimentary petrography when advocating the origin of certain sandstones, M.N. Bramlette (1934) and R. G. Tickell (1924) introduced the correlation of strata using heavy-mineral compositions, and G. Rittenhouse (1943) made an excellent contribution to the study of the behaviour of heavy minerals. Furthermore, in the United States, the study of the textures of sediments and sedimentary rocks commenced in this period (*see* Chapter 6.1.3). A text-book by Krumbein and Pettijohn (1938) was a benchmark publication, which, more than sixty years on, still remains a major reference work for anyone studying sedimentary rocks.

### 4.2.2 Sedimentary petrology

Sedimentary petrology is the scientific study of sedimentary rocks using *all* the features found in sediments and sedimentary rocks. It differs, therefore, from sedimentary petrography, which lays emphasis on the description of the petrographical characteristics of sedimentary rocks. In the United States there was a period from the 1930s to the 1960s, when 'sedimentary petrology' and 'sedimentary mineralogy' implied 'sedimentology' (Twenhofel, 1941).

It was after World War II (around 1945) that the field of sedimentary petrology made remarkable progress. Paul D. Krynine (1902–1964) (Fig. 4.2), a professor

**Fig. 4.2** Paul D. Krynine (Williams, 1965). (*Permission Geological Society of America.*)

of Pennsylvania State University, in particular, contributed much towards the modernization of sedimentary petrology, and it was Prof. F. J. Pettijohn, who elevated this discipline into a mature science.

Petrological studies of sedimentary rocks follow quite different methods depending upon their clastic or non-clastic nature, as shown by Pettijohn *et al.* (1973), Pettijohn (1975), Mizutani *et al.* (1987), Boggs (1992) and others. Here, the study of sandstones is used to illustrate some of the main achievements in the petrological studies of clastic sediments.

### 4.2.3 Modernisation of sedimentary petrology as illustrated in sandstone classification

P. D. Krynine was the first to create a dynamic discipline for sedimentary petrology through the discussion of earth history, based on tectonic and climatic influences on the mineral compositions of sedimentary rocks using the sedimentary petrography approach established by H. C. Sorby (Krynine, 1946a, b). Krynine systematically classified clastic rocks and discussed their geological significance by relating these classifications to tectonic movements (Krynine, 1941a, 1942, 1948). Krynine thus established the following scheme on the basis of his Appalachian model:

Pre-geosynclinal sediments (peneplanation stage): typically orthoquartzite;

Geosynclinal sediments (basinal stage): typically greywacke;

Post-geosynclinal sediments (post-orogenic stage with crustal uplifting): typically arkose.

Here orthoquartzite is a sandstone with more than 95% quartz grains as the framework fraction, which are generally well-sorted, well-rounded and cemented with secondary silica; greywacke is a dark grey ill-sorted sandstone with more than 25% rock fragments and a significant amount of clay matrix; and arkose is a sandstone containing more than 25% feldspar which exceeds the percentage of rock fragments.

Orogenic cycles and their correlation with sedimentary characteristics formulated by Krynine exerted a significant influence upon geology at that time (Kay, 1951). His scheme was refined and systematized by Pettijohn (1949, 1957a, 1975) leading to a widely accepted methodology for the study of sediments.

Sedimentary petrology, as modernized by Krynine, has developed mainly in the study of the mineral compositions of clastics. An unavoidable problem in this process, however, arose in the classification of sandstones. In the two decades following the proposal for sandstone classification by Krynine in 1948, and as the study of sandstones progressed, more than fifty revisions of the sandstone classification were proposed in more than ten countries and in as many as seven languages! The study of sandstones was thrown into confusion. The largest cause of this confusion arose from the genetic classification of sandstones.

The basic characteristics of sandstones are formed by texture and mineral composition which, in turn, are controlled by:

(1) provenance factors (Pettijohn, 1954),

(2) fluidity factors (Pettijohn, 1954),

(3) textural maturity (Folk, 1951),

(4) mineralogical maturity (Folk, 1954; Pettijohn, 1954),

(5) diastrophism (Krynine, 1948),

(6) primary sedimentary structure (Packham, 1954; Crook, 1960),

(7) climate and weathering (Strakhov, 1962), and

(8) post-depositional alteration (Cummins, 1962).

If we take genesis into consideration as a classification factor then it is necessary to reflect as many as possible of these factors. Because no classification incorporates all of these criteria within a single framework, each author, depending on his priorities, combined them in different ways.

In his prototype of modern sandstone classification, Krynine (1948) argued for a close relationship between tectonic cycles and classes of sandstone – the concept of the tectonic control of sedimentation. Thus, Krynine (1941a,b,c,d,e, 1942, 1948, 1951) proposed correlations between tectonic stages and sandstone types, namely: the peneplanation stage = orthoquartzite; the geosynclinal stage = greywacke; and the post-orogenic stage = arkose. This notion of three basic sandstone clans (orthoquartzite, greywacke, and arkose) seems to have been long established in the geological literature. In consequence, it had been generally accepted that rock names indicate the origin of rocks and their depositional environments, as, for example, greywackes imply either geosynclines or environments of flysch deposition. Worse, the confusion of definitions of greywacke was serious, as more than thirty definitions had been proposed since 1948. In such a situation, it had become clear that the so-called type greywacke of the Harz Mountain in Germany and the type arkose of the Auvergne Mountains in France overlapped each other in composition (Helmbold and van Houten, 1958; Huckenholz, 1963). Naturally, the definitions of greywacke and arkose became blurred and caused confusion.

**Fig. 4.3** Robert H. Dott, Jr. (photo by H. Okada).

In order to remove uncertainties and ambiguities attached to these terms, the total abandonment of greywacke, arkose and orthoquartzite was advocated. As a result, new schemes of descriptive classification based on observable and measurable parameters were proposed by Robert H. Dott, Jr. (Fig. 4.3), professor of the University of Wisconsin (Dott, 1964) and H. Okada (1971a, b). These classification schemes are constructed both on the texture of sandstones (mean grain size and clay-matrix content) and on the composition of the detrital framework constituents. For example, the classification of Okada (1968a, 1971a, b) is threefold in its organization: (a) grain size, (b) clay matrix and cement, and (c) composition of framework constituents (Fig. 4.4).

This threefold classificatory scheme of sandstones is as follows:

(1) Sandstones are divided into 'wacke' and 'arenite', with more than and less than 15% clay matrix respectively. As such, the nature of sandstones is defined by the clay-matrix content.

(2) When sandstones have little clay matrix and are rich in chemical cements and non-terrigenous components such as allochemicals and authigenic minerals (glauconite, siderite, phosphate, etc.), adjectival modifiers such as

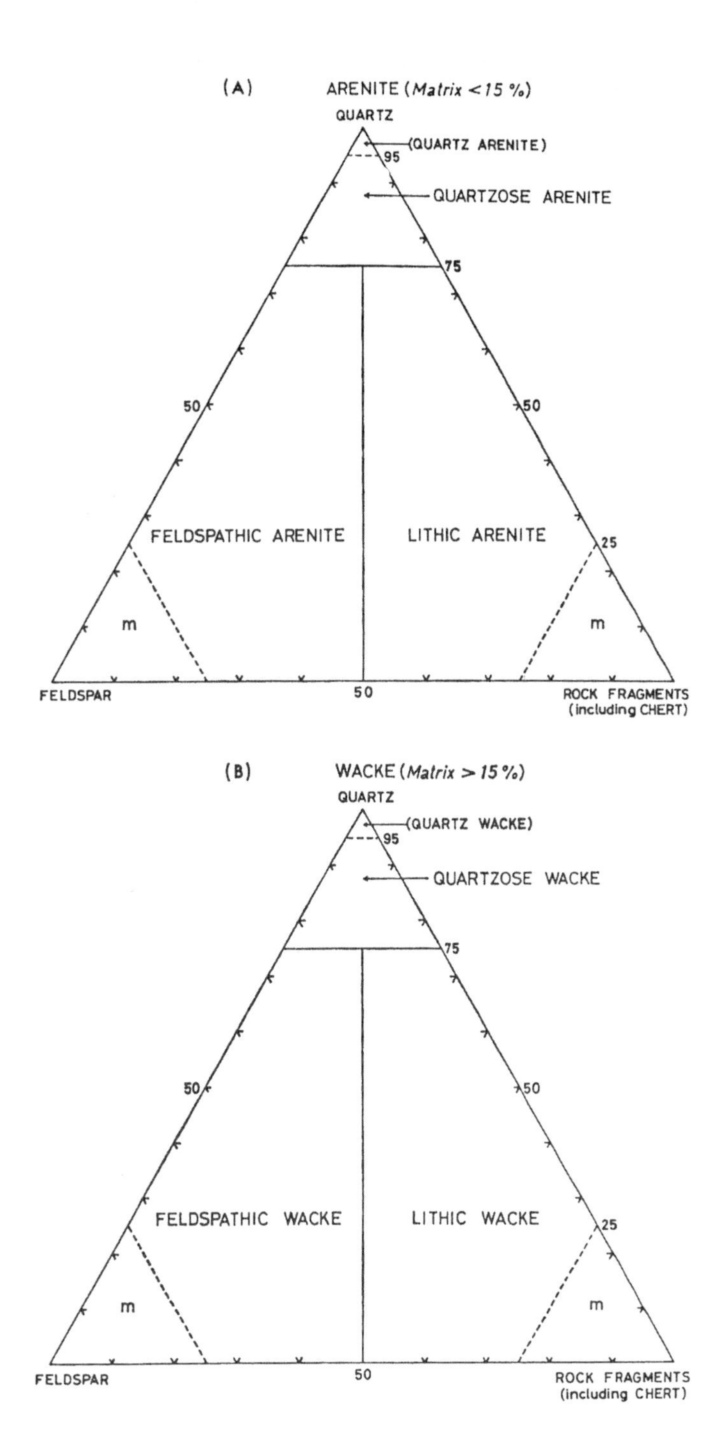


**Fig. 4.4** Okada's scheme of sandstone classification (Okada, 1971a, b). m indicates the monomineralic field. For more detail, see the text.

'calcareous', 'siliceous', 'zeolitic', 'glauconitic', etc. are added to the initial of rock names. Indurated sandstones generally show that the contents of the matrix and the framework constituents indicate considerable variation as the mean grain size changes, though trends of those variations are not the same in all cases. Thus, adjectives such as 'coarse-grained', 'medium-grained', and 'fine-grained' are also added.

(3) The use of triangular diagrams for sandstone classification is effective and convenient for laboratory study because sandstone compositions can be indicated precisely in this way. All framework grains are grouped into three end-members: quartz (Q), feldspar (F), and rock fragments (R) (or lithic fragments (L)). Chert is grouped with rock fragments (not with quartz) because chert grains behave like other rock fragments. The classification triangle with Q : F : R (L) is subdivided into three major fields: quartzose arenite or wacke, feldspathic arenite or wacke, and lithic arenite or wacke (Fig. 4.4). The quartz arenite or wacke field is retained for sandstones with more than 95% quartz. Similarly, fields with more than 75% feldspar and rock fragments, repectively, are for practical purposes designated as 'monomineralic' (m in Fig. 4.4). Sandstones in monomineralic fields are named by replacing 'feldspathic' or 'lithic' with a specific term for the most predominant mineral or rock-clast; for instance, 'serpentine arenite' (Okada, 1964; Zimmerle, 1968), 'plagioclase arenite' or 'plagioarenite' (Okada and Nakao, 1968; Okada, 1973), volcanic wacke (Crook, 1960; Chappel, 1968).

In addition, as a descriptive classification, a symbolic or numerical classification of sandstones has been proposed. For example, a sandstone composed of 50% quartz, 20% feldspar, and 30% rock fragments is shown as $Q_{50}F_{20}L_{30}$ (Boggs, 1967).

### 4.2.4 Sandstone study and its significance in the reconstruction of provenances and tectonic environments

Even if the sandstone classification itself is not the purpose of study, petrological studies of sandstones in relation to their classification have not only promoted clarification of petrological features of sandstones, but have also encouraged studies of provenances and tectonic environments based on the compositions of sandstones. Representative of this trend in the 1970s, William R. Dickinson (Fig. 4.5), a professor of Stanford University, developed a new means of determining the

**Fig. 4.5** William R. Dickinson (photo by H. Okada).

quantitative detrital modes of sandstones by dividing sand grains into constitutive mineral species (Dickinson, 1970, 1985; Dickinson and Suczek, 1979; Dickinson *et al.*, 1983). This approach gave an impetus to a new development of sandstone studies (Ingersoll, 1983; Zuffa, 1985).

Following the procedure of Dickinson, constituent particles of sandstones are separated into the mineral species indicated in Table 4.1, and their compositional fields are plotted on triangular diagrams indicative of sand derivation from different types of provenances (Fig. 4.6). One of the key points is that lithic fragments are classified into mineral species. It should be noted, however, that

**Table 4.1** Classification and symbols of grain types (modified from Dickinson, 1984).

| | | |
|---|---|---|
| **A** | Quartzose grains | Qt = total quartzose grains |
| | Qt = Qm + Qp | Qm = monocrystalline quartz (> 0.625 mm) |
| | | Qp = polycrystalline quartz |
| **B** | Feldspar grains | F = total feldspar grains |
| | F = P + K | P = plagioclase grains |
| | | K = K-feldspar (potassium feldspar) grains |
| **C** | Unstable lithic fragments | L = total unstable lithic fragments |
| | L = Lv + Ls | Lv = volcanic/metavolcanic lithic fragments |
| | | Ls = sedimentary/metasedimentary lithic fragments |
| **D** | Total lithic fragments | Lt = total polycrystalline lithic fragments |
| | Lt = L + Qp | |

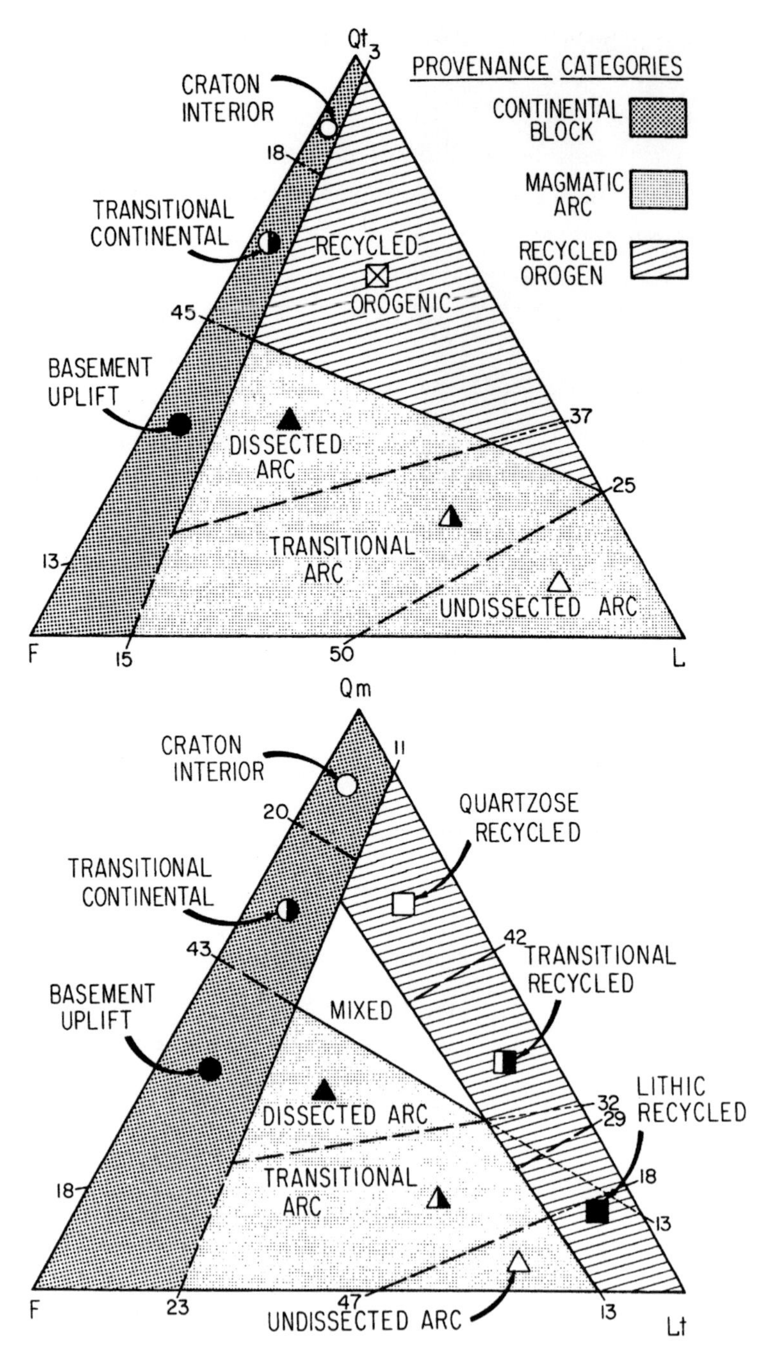


**Fig. 4.6** Dickinson's diagram showing sand derivation (Dickinson *et al.*, 1983). Qt: total quartzose grains, Qm: total monocrystalline quartz, F: total feldspar grains, L: total unstable lithic fragments, Lt: L + Qp (Qp: polycrystalline quartz). (*Reproduced by permission of the Geological Society of America.*)

this approach is presented to infer sandstone provenance but not to describe the petrological features of sandstones.

Quite separately in the 1980s, chemical methods of analysing the tectonic environments of sandstones were developed. For instance, in the chemical composition of sandstones, Bhatia (1983) distinguished four tectonic environments, namely inactive marginal areas, active marginal areas, continental arc areas, and oceanic arc areas, on the basis of compositional relations of $TiO_2$, $Al_2O_3$ / $SiO_2$, $K_2O$ / NaO, $Al_2O_3$ / (CaO + $Na_2O$), and $Fe_2O_3$ + MgO. Roser and Korsch (1986) convincingly distinguished the three environments of inactive margin, active margin and oceanic arc regions on the basis of the relation between the values of $K_2O$ / $Na_2O$ and $SiO_2$. Another interesting method was proposed by Kiminami *et al.* (2000) in order to distinguish tectonic environments of sandstone deposition on the basis of the chemical composition of clastic sediments. It is based upon the newly proposed Basicity Index (= (FeO + MgO) / ($SiO_2$ + $K_2O$ + $Na_2O$) and $Al_2O_3$ / $SiO_2$ (BI diagram) (Fig. 4.7). Kiminami *et al.* successfully discriminated between the immature arc, the dissected arc, and the mature island arc within island arc environments.

Other methods worthy of mention are the studies of the source rocks of clastics based on particular minerals such as garnet (Morton, 1985; Takeuchi, 2000)

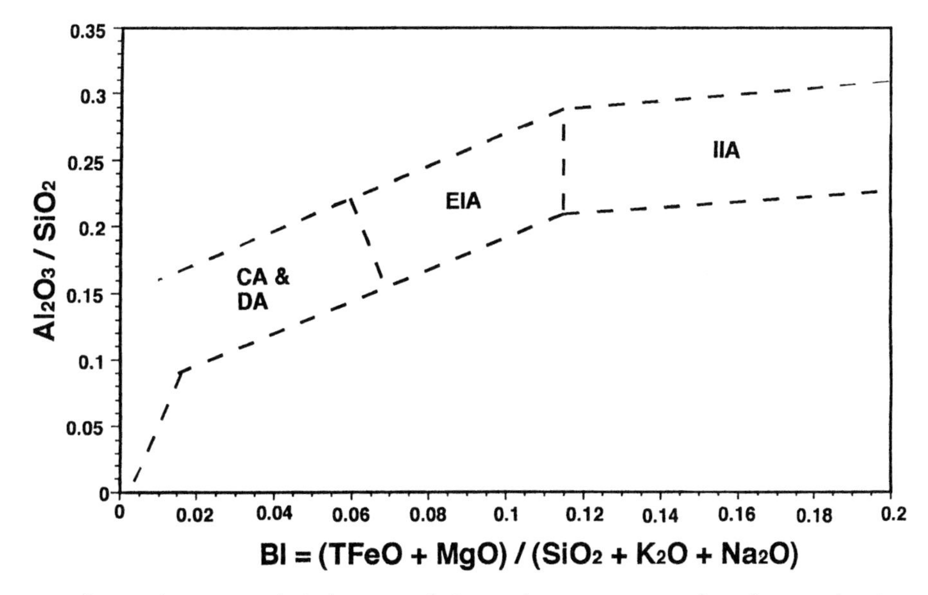


Fig. 4.7 BI (Basicity Index) diagram of chemical compositions of sandstones showing discrimination of arc settings (revised from Kiminami *et al.*, 2000). IIA: immature island arc, EIA: evolved island arc, CA: continental arc, DA: dissected arc.

and spinel (Arai and Okada, 1991; Hisada and Arai, 1993; Hisada *et al.*, 1999). Such studies have been accelerated by the development and spread of advanced instruments for chemical analysis such as electron probe microanalyser (EPMA).

Furthermore, as described below, there have been several proposals for stratigraphic and earth-historical evaluations of the petrological features of sandstones, concepts of petrofacies, petrographic zone and petroprovince.

Petrofacies (Dickinson and Rich, 1972) is a stratigraphic unit characterized by microscopic features of sandstones that can be applied to the correlation of strata within a sedimentary basin. Petrographic zone (Okada, 1977) is also a stratigraphic unit, this time characterized by particular petrological components and it is useful as a chrono-stratigraphic indicator. Okada (1977) showed that the sandstones of the Cretaceous to Palaeogene Shimanto Supergroup in Kyushu can be subdivided into three petrographic zones, using the feldspar content. A petroprovince (Okada, 1997) is defined as a sedimentary province characterized by a group of sediments with a unity of age, lithofacies, petrological features and distribution. Okada (1997) discriminated three petroprovinces for the Cretaceous in the Japanese Islands, that is, the Nemuro Petroprovince characterized by andesitic volcaniclasts, the Red-bed Petroprovince of the Inner Zone of Southwest Japan, and the Toyonishi – Kawaguchi Petroprovince characterized by quartz arenites.

Furthermore, the use of mass spectrometry is also important in sedimentary petrology for determining the age of single clasts or mineral grains (Hole, 1998). Dating these means that both the nature of the source rock and its uplift history can be determined. Thus, a key link between tectonics and sedimentation is established. Good examples of such studies were presented recently by Longman *et al.* (1979) and Kelley and Bluck (1989). Longman *et al.* (1979) clarified the source of the Upper Dalradian (Ordovician) clastics in Scotland by Rb-Sr analytical data. Kelley and Bluck (1989) determined detrital mica ages in the Southern Uplands of Scotland using a $^{40}Ar$–$^{39}Ar$ laser probe and indicated the source of detrital volcanic clasts.

## 4.3 Texture of sediments

The study of the texture of sediments requires the observation and analysis of the morphology of the constituent grains of sediments, as well as of the sedimentary texture controlled by the space relationships between particles. This method is a fundamental procedure for describing the properties of particles and sedimentary rocks, and has a long history as a basic field of sedimentary petrography. There is no single word to describe the study of sediment texture, but it is normally referred to as the 'Properties of Sedimentary Rocks' (Folk, 1968), the 'Texture of Sediments' (Pettijohn, 1975), the 'Properties of Sedimentary Particles' (Friedman and Sanders, 1978), 'Sedimentary Textures' (Boggs, 1986, 1995), and so on.

Factors characterizing sediments are controlled by the individual constituent particles. They reflect features of source rocks, weathering processes and processes of transportation and deposition, and thus can be said to be an essential indicator of sedimentary environments. For analysing the characteristics of the constituent particles, systematic observations and measurements of (1) grain size, (2) shape, (3) surface texture, and (4) mineral composition must be taken into account. Working methods were developed in the United States mainly to meet the needs for prospecting oil and gas resources (*see* Chapter 6.1.3), American School). The greatest contributor to the synthesis of these methods was W. C. Krumbein (1902–1975) (Fig. 4.8), who was a student of Prof. F. J. Pettijohn at Chicago University and became a professor at Northwestern University. Both Krumbein and Pettijohn also rendered a great service by introducing quantitative methods

**Fig. 4.8** William C. Krumbein (Howland, 1975).
(*Reproduced by kind permission of the University of Chicago Press.*)

and establishing mathematical and statistical approaches to the study of sediments (Krumbein and Pettijohn, 1938). Their work became a major force in creating the establishment of systematised sedimentology.

In the study of the texture of sediments, the technique dealing with the constitution of particles in sediments is called *granulometry*, and that studying the shape characteristics of them is called *grain morphometry* (Köster, 1964).

### 4.3.1 Granulometry

For this purpose the following techniques are widely adopted: (1) sieving, (2) settling, and (3) observational methods including microscopic analysis. The sieving technique is useful for grain-size measurements of coarser-grained sediments ranging from granule to fine sand, while a settling technique is applied to finer-grained sediments such as silt and clay particles. Observational methods include direct measurements of grain size with a magnifying hand lens and more sophisticated methods using the microscope and electron microscope. The last are generally applied to size analysis of clay and consolidated sediments while the first two are only applicable to unconsolidated sediments. A large number of techniques and their evaluations have been proposed for grain-size analyses, as explained in Krumbein and Pettijohn (1938), Folk (1966), Pettijohn (1975), Boggs (1992), Sengupta (1994) and by many others. It is worthy of special mention that Gerald M. Friedman (Fig. 4.9), professor of Rensselaer Polytechnic Institute, New York, proposed an important method to carry out grain-size analyses of consolidated sandstones in thin sections under the microscope (Friedman, 1958). His conversion diagram from the thin-section data of grain-size analysis to that of sieving analysis is particularly useful (Friedman, 1958, 1962). A mathematical model for comparison between sieve, settling and thin-section data was also proposed by Burger and Skala (1976).

### 4.3.2 Grain morphometry

Methods for analysing particle shape and its interpretation are called grain morphometry (Köster, 1964; Pryor, 1971; Brock, 1974). The particle shape can be expressed by form, roundness, and surface texture. The form is generally related to sphericity. As a general description of these morphological factors, a concept of pivotability or rollability was proposed by F. P. Shepard and R. Young (1961).

It was Hakon Wadell (1932a), who first introduced a quantitative method to express the shape of clastic particles. His method was to imagine three axes (long, medium, and short) crossing at a right angle at the centre of a particle,

**Fig. 4.9** Gerald M. Friedman (courtesy G. M. Friedman, 2000).

and to express the shape and sphericity as a ratio of their relative lengths. This idea was further developed by Zingg (1935), Krumbein (1941a) and Sneed and Folk (1958). Among them, the sphericity diagrams of Zingg and Sneed and Folk, which define the basic shape of particles as equant or spherical, prolate or roller, oblate or disk, and bladed, are still widely used. Such particle shapes reflect, on the whole, the original forms of minerals and lithic fragments and tend to be influenced more intensely in processes of transportation and deposition. They are more obvious in the particles with larger sizes. Such particle shapes exert a major influence on their sedimentation rates.

In general, perfectly spherical particles are controlled by the Stokes and Impact Laws in their sedimentation velocity. That is to say, the sedimentation velocity of particles smaller than 0.1 mm in diameter follows the Stokes Law, which states that the sedimentation settling velocity is proportional to the square of diameter (Stokes, 1851), whereas the sedimentation velocity of particles larger than 1 mm in diameter follows the Impact Law, which defines that the sedimentation velocity

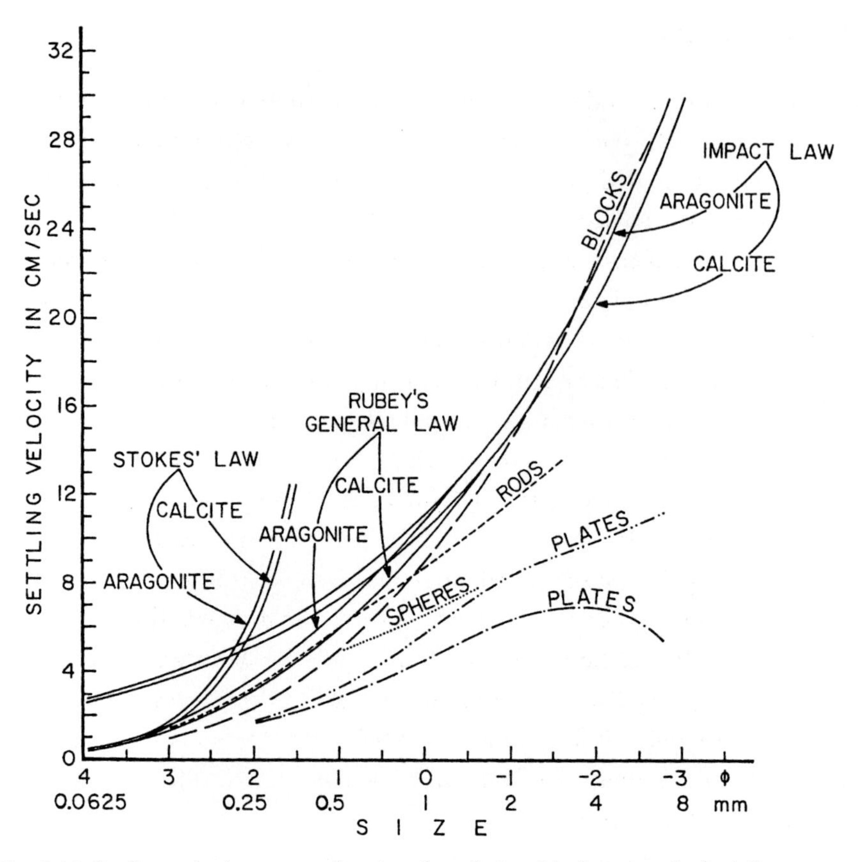


**Fig. 4.10** Settling velocity curves showing the relationship between Stokes', Impact, and Rubey's Laws (after Maiklem, 1968). Aragonite (specific gravity 2.95) and calcite (s.g. 2.72) grains were used for settling-velocity experiment in sea water (density 1.03, temperature 23.5). (*Reproduced by permission of the International Association of Sedimentologists.*)

of particles is controlled by the mass of the particles and the viscosity of liquids (Rubey, 1933). In addition, particles between 0.1 mm and 1 mm in size follow the Rubey Law (Rubey, 1933) (Fig. 4.10). However, particles with irregular shapes show unpredictable behaviours in their sedimentation. When trying to use the Stokes–Impact Law, W. R. Maiklem (1968) showed how the sedimentation velocity of particles with such different shapes as block, rod, plate, and sphere deviate from the curve predicted by the Stokes–Rubey–Impact Laws (Fig. 4.10). This diagram clearly suggests that coin-shaped or leaf-like particles such as micas and shale clasts show complicated sedimentation behaviours that reduce their settling velocities.

### 4.3.3 Roundness

The roundness of individual particles supplies important information reflecting the mineral composition of particles, their origins, their size, transportation process, and transported distance. Pioneer works on the quantitative estimation of roundness of particles were made by Wadell (1932a, 1933b) and Zingg (1935). The method of evaluation of roundness proposed by Wadell was to give the ratio of the average radius of curvature of observable corners or edges to the radius of curvature of the maximum inscribed sphere. Since this method is, in practice, difficult to apply, a silhouette or a visual comparison method was proposed as a quick and practical determination of roundness. This method is well known as the Powers' Roundness for sands (Powers, 1953) and as the Krumbein's Roundness Image for gravels (Krumbein, 1941a), both of which are popular among workers.

M. C. Powers (1953) divided the roundness into six stages; very angular, angular, sub-angular, sub-rounded, rounded, and well-rounded. Folk (1955) revised this roundness classification quantitatively and proposed the *rho* ($\rho$) scale. According to Folk's method, the Powers' six stages of roundness are expressed as 0.0–1.0 for very angular, 1.0–2.0 for angular, 2.0–3.0 for subangular, 3.0–4.0 for subrounded, 4.0–5.0 for rounded, and 5.0–6.0 for well-rounded. This scale indicates that the roundness of a complete sphere is 6.0 $\rho$ whilst that of usual sand grains is 2.5 $\rho$ (subangular).

It is said that experimental abrasion was carried out first by a French geologist, A. Daubrée (1879). It was further developed by Wentworth (1919), Krumbein (1941b), Kuenen (1956), and Bradley (1970) with variable results based upon experimental observations. As far as the available data are concerned, it is generalized that hard and highly resistant particles like quartz and zircon are hardly abraded, and that larger sized particles such as gravels are more easily worn than sand-sized ones. In particular, sand particles less than 0.1 mm in diameter show almost no abrasion during transportation in water. However, a mechanism by which sand particles are abraded, is caused by direct collisions even of fine grains during aeolian transportation. This is why desert sands tend to show well-rounded features, i.e. 'millet-seed' sandstone (Jukes-Browne, 1922, p. 222).

When particles move down a slope under the influence of gravity, the rate of movement of particles depends upon their shape. Such shape characteristics that control the movement are defined as rollability (Shepard and Young, 1961). Kuenen (1956, 1963) also inspected the rollability of particles by experiments and field observations, emphasizing its importance.

### 4.3.4 Surface texture

The physical and chemical processes that operated on particles during transportation and deposition are recorded as their surface texture, an attribute as important as roundness and sphericity. The characteristics of surface texture of particles show the distinct history of each particle, and are useful as effective indicators of depositional environments. This is well illustrated by striated pebbles and boulders which usually indicate a glacial environment.

The grain surface textures are divided into two types of small-scale relief; one is of physical origin, shown by polished surface, frosted glass-like surface, striated patterns, conchoidal breakage, arc-shaped steps, small-scale indentations and small-scale ridges, whilst the other is of chemical origin, indicated by wavy, etched, and low-relief solution surfaces. These textures are variable in size from directly visual to microscopic observations, and for quartz sand grains electron microscopy has been applied effectively. D. H. Krinsley, a professor of New York State University, and his colleagues, first systematically applied the scanning electron microscope (SEM) to the study of quartz sand grain surface textures (Krinsley and Takahashi, 1962a, b, c). They distinguished the surface textures of quartz sand grains using high resolution SEM images (Krinsley and Doornkamp, 1973), and discussed their origin and environmental significance (Krinsley and Smalley, 1972, 1973) (Fig. 4.11). It should be borne in mind, however, that when discussing sedimentary environments on the basis of surface textures of grains, there are strong possibilities that similar textures may have been produced by different processes.

### 4.3.5 Fabric

Fabric is an expression of the structure of sediments and sedimentary rocks that indicates the spatial relation of constituent particles and their orientations. It is generally expressed by orientation and packing. The data of fabric in sedimentary geology have been quite useful for basin analysis and analysing diagenesis (*see* Chapter 3: Palaeocurrent analysis and basin analysis). The orientation of particles is determined by that of the long axis of particles (a-axis), directions of the inclined surface of the a-b axes or imbrication, and so on. Furthermore, fabric is defined in two ways: depositional fabric, which is the fabric produced at the time of deposition and reflects sedimentation conditions; and deformation fabric, which is produced by deformation after deposition but before lithification. Normally, tills of glacial origin appear structureless and devoid of bedding and

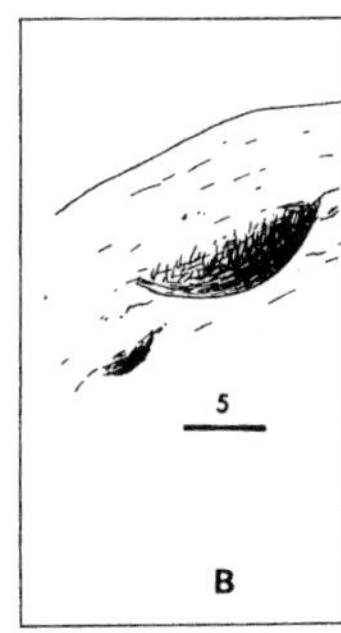

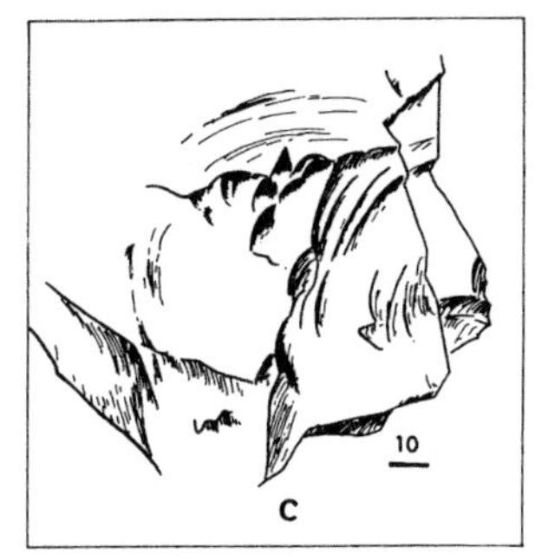

**Fig. 4.11** Surface textures of quartz sand grains (sketched from SEM images in Krinsley and Margolis, 1969; Krinsley and Smalley, 1972). **A**: Irregular, partially oriented V-shaped depressions most characteristic of sand grains from all aqueous environments – rivers, beaches, continental shelves, and deep-ocean basins. **B**: Polished and frosted surface of a well-rounded desert sand, showing wind action. **C**: Conchoidal fractures with sharp edges produced by grinding and breaking due to glacial action. **D**: Portion of a sand grain showing evidence of two episodes of mechanical and chemical actions. The left-hand side indicates irregular conchoidal breakage patterns by glacial action, which are subsequently modified to V-shaped patterns (right-hand side) in subaqueous environment. Scales in all figures in microns. (*Reproduced by permission of the New York Academy of Sciences.*)

internal orientation. Hugh Miller (1884) showed interest in the systematic study of till fabric. It was a very early observation of the orientation of clasts. Krumbein (1939) and Rusnak (1957) discussed the significance of grain orientation.

Important depositional fabrics are the grain-supported and the matrix-supported fabric. In the former, grains support themselves by direct contact with each other, while in the latter, larger grains float within muddy fine-grained materials without direct contact between the larger grains. In general, gravel deposits are controlled by different processes in transportation and deposition, depending upon the difference in grain-supporting fabrics. Gravels, which are transported by rolling on the bottom by traction force, as in river sand bars and foreshore, show a grain-supported fabric. In contrast, when matrix-forming fine-grained materials are transported simultaneously with gravels and are deposited rapidly, a matrix-supported fabric is produced, as represented by pebbly mudstone and debris-flow deposits, which are the deposits of a variety of sediment gravity flows. It is generally said, but not always true, that these kinds of sediments show no orientation. Actually, however, there are many examples where the gravels are oriented with a parallel arrangement of the a-axis by mutual collision during transportation (Okada and Tandon, 1984). Walker (1975) also pointed out that resedimented gravels are oriented with the a-axis parallel to the current direction and show an oriented fabric with imbrication of the a-b plane of gravels inclined upcurrent.

Packing is the density of particle arrangement in a limited space. In general, the packing of clastic sediments and sedimentary rocks is closely related to porosity. Graton and Fraser (1935) showed the packing characteristics of solid spheres of uniform size range from the tightest rhombohedral packing with a porosity of 25.95% to the loosest cubic packing with a 47.64% porosity. Rittenhouse (1971) and Beard and Weyl (1973) discussed changes of the porosity of sands in the process of diagenesis in addition to changes of permeability and bulk density. In particular, since Middleton and Hampton (1973) proposed the concept of liquefied sediment flow or fluidized sediment flow as one of the sediment gravity flow processes, more attention has been paid to liquefaction or fluidization of unconsolidated sediments with rapid changes in porosity and to the phenomenon that water-saturated sandy sediments get liquefied by the increase in intergranular pore water pressure (Lowe, 1975). Such studies of the texture of sediments have also contributed much fundamental information in exploring for oil and gas accumulations.

### 4.3.6 The significance of sedimentary textures: the maturity concept

The concept of maturity, which indicates evolving characteristics of clastic sedimentary textures, is an important indicator for showing the dynamic environments of sediments. W. J. Plumley (1948), a student of Prof. F. J. Pettijohn at the University of Chicago, first introduced this concept in the study of sediments. Plumley, associating maturity with roundness, sphericity, and rock features of clastic particles, discriminated between mature sediments with larger values of roundness and sphericity and with small amounts of soft rock fragments, and immature sediments with opposite features. The maturity concept was enhanced by the establishment of the concept of textural maturity by Robert L. Folk (Fig. 4.12), a professor of the University of Texas (Folk, 1951). Folk brought textural and compositional maturity together in a brilliant unifying synthesis. According to him, the texture of sediments is the product of the mechanical energy involved during transportation and deposition thus reflecting the physical nature of transportational and depositional environments. He claimed that clastic sediments decrease in clay matrix and increase in sorting and roundness of the sand grains as a response to increased energy levels in the processes of weathering,

**Fig. 4.12** Robert L. Folk (photo by H. Okada).

transportation and deposition, and that sediments show a sequential development in textural maturity from an immature stage, sub-mature stage, mature stage, and finally to a super-mature stage (Fig. 4.13) .

On the other hand, F. J. Pettijohn (1949) proposed the concept of mineralogical maturity. This is intended to express the maturity of sedimentary rocks based on the stability of rock-forming minerals. For example, such measures as quartz / feldspars ratio, (quartz + chert) / feldspars ratio, and (quartz + chert) / (feldspars + rock fragments) ratio are adopted. Pettijohn emphasized time or duration of action in the concept of mineralogical maturity. The concept of mineralogical maturity was applied to the heavy mineral components of terrigenous sediments. As some heavy minerals are selectively destroyed during transportation and diagenesis, only physically and chemically stable minerals constitute the remaining heavy minerals, which are characteristic of mature sediments. In this respect, sediments of high maturity contain larger amounts of such stable minerals as zircon, tourmaline, and rutile. The combined percentage of these minerals was defined as the ZTR index by J. F. Hubert (1962). In addition, Y. Sato (1969) advocated the

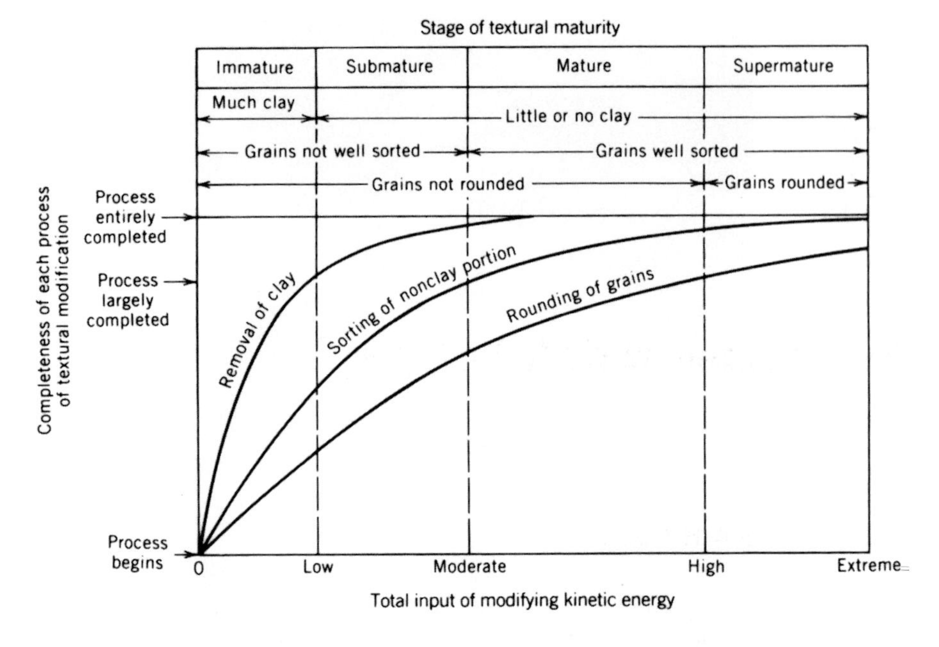


**Fig. 4.13** Folk's diagram showing the concept of textural maturity (Folk, 1951).

importance of garnet as one of the index minerals of mineralogical maturity and introduced the ZGT ratio for characterizing local geological features in Japan.

Another attempt at characterising the stability of sedimentary rocks and minerals is the concept of chemical maturity. It is defined, for example, as $Al_2O_3$ / $Na_2O$, $Al_2O_3$ / $SiO_2$, $K_2O$ / $Na_2O$, $K_2O$ / ($Na_2O$ + CaO), and so on. L. Harnois (1988) introduced a chemical index of weathering (CIW) as 100 ($Al_2O_3$) / ($Al_2O_3$ + CaO + $Na_2O$ + $K_2O$). Combining mineralogical maturity and chemical maturity is called *compositional maturity*.

As stated above, as studies such as mineral stability series in weathering (Goldich, 1938), relative abrasion resistance of minerals (Thiel, 1940), and chemical stability of minerals (Pettijohn, 1941; Smithson, 1941; Reiche, 1943) developed, the maturity concept of sediments became an integral part of clastic sediment studies.

# 5

# Development of lithology

## 5.1 Facies concept

Perhaps no other geological term has been subjected to such a variety of interpretations as the word 'facies'. The term *facies* is a Latin word meaning face, figure, appearance, state, look, aspect, and condition. It is said that Nicolaus Steno (1669) (*see* p. 5) introduced the word into the geological vocabulary. By using it, he meant to express all the features of a part of the earth's surface during a certain geological time. The meaning of '*facies*' as generally used in sedimentology can be expressed as a combination of the lithological and the palaeontological aspects of sediments and sedimentary rocks at a particular stratigraphic position.

The present concept of 'facies' dates back to the stratigraphic study of the Upper Jurassic in the Jura Mountains of France and Switzerland by Amand Gressly (1814–1865). Gressly recognized a close relationship between lithological features and the contained fossils. He differentiated the Jurassic sedimentary facies into a reef facies and a muddy facies, and suggested that the former implied a shore to shallow-sea environment while the latter reflected a quiet, deep-sea environment. Thus, Gressly showed the recognition of facies to be a valuable tool for the reconstruction of sedimentary environments. He pointed out that facies changes are caused by changes in depositional environments, and summarized his conclusions in five facies laws (Gressly, 1838).

> **First law:** Each facies of a stratigraphic unit presents very distinct lithological or palaeontological characteristics.
>
> **Second law:** In different stratigraphical units the same lithological facies affects the palaeontological characteristics in much the same way.

**Third law:** Different facies may have sharp boundaries in a horizontal direction, but sometimes they pass into each other through transitional (intermediate) varieties.

**Fourth law:** In the stratigraphic section the diversity of facies increases from the bottom to the top.

**Fifth law:** The diversity of facies may become greater or less in different areas.

Gressly's concept of facies was firmly rooted in geological field studies. Based on his observations, it is clear that the understanding of the concept of facies is indispensable in the study of sedimentary rocks. Curt Teichert (1958) investigated in detail the history and regional recognitions of the concept of facies. Before Teichert and continuing to the present, the concept of facies has been used with a large variety of meanings; for example, *Faziesbezirk* (facies tract) by Johannes Walther (Walther, 1894), *Faziesreihen* (facies sequence) as vertical facies changes (Walther, 1894; Frank, 1936), metamorphic facies produced under the same metamorphic conditions (Eskola, 1922), tectonic facies related to tectonic movements, a collective term for rocks that owe their present characteristics mainly to tectonic activity and different from tectofacies (Sander, 1912), strain facies (Hansen, 1971), fabric facies as the type of deformation lamella in quartz grains in metamorphic rocks (Hara *et al.*, 1976), deformation facies indicating the mode and geometry of deformation (Uemura, 1981), granulometric facies showing sedimentary characteristics on grain-size diagrams (Rivière, 1952), tectofacies is interpreted tectonically (Sloss *et al.*, 1949), orogenic stage facies showing the synchroneity of orogenesis (Kobayashi, 1956), geosynclinal facies (Aubouin, 1965), turbidite facies and flysch facies (Dzulynski and Walton, 1965), lithofacies, biofacies, elemental facies based on isotope characteristics of sedimentary rocks (Adams and Weaver, 1958), geochemical facies showing sediments deposited under similar chemical environments (Pustovalov, 1933; Adams and Weaver, 1958), microfacies indicating microscopic litho- and biofacies (Cuvillier and Sacals, 1951), ichnofacies relating to trace fossils (Seilacher, 1967, 1978; Ekdale *et al.*, 1984), and so on.

Johannes Walther (1893, 1894), who greatly influenced the development of the facies concept, considered that the facies of sediments reflected the condition at the time of sedimentation. This view has been accepted by many workers. He further called a lithofacies combination a *Faziesreihen* (facies sequence) when lithofacies differed from each other but were produced under a related set of conditions (Walther, 1894).

Teichert (1958) summarized the facies concept as follows:

> The term facies preferably should be restricted to sedimentary rocks. Facies is the sum total of all primary characteristics of a sedimentary rock, from which the environment of its deposition may be deduced. Facies should be used as a strictly descriptive term, referring to the primary lithologic and/or palaeontologic characteristics of a sedimentary rock. Environmental interpretation is the ultimate aim of facies studies, but environment is not facies. It produces facies. Interpretations may be subject to change; a facies which indicates a littoral environment to one observer, may suggest a continental (aeolian) environment to another. Hence, terms like littoral and aeolian facies are meaningless, or at least ambiguous and subject to misinterpretation.

Actually, however, the facies concept with genetic implications is used widely and frequently.

The term 'sedimentary facies' was proposed by R. C. Moore (1949). It means that a particular stratigraphic unit is characterized by features that clearly distinguish it from other units, which, in turn, are restricted to particular conditions. Moore separated sedimentary facies from lithofacies. He regarded the lithofacies as the record of sedimentary environments that show physical and chemical characteristics, whereas biofacies indicates biological aspects.

Moore also defined the concept of facies as the expression of the collective characteristics of sedimentary rocks, which include both descriptive properties such as lithological and palaeontological aspects and interpretative concepts suggesting sedimentary environments. Therefore, it may be said that a certain lithofacies is defined either as the sandstone facies or as the fluvial facies. The former is a descriptive expression, and the latter is an indication of sedimentary environments based on interpretation. As stated above, it may be more convenient to follow either of the expressions, depending upon the purpose of description. As R. Anderton (1985) pointed out, a certain expression may become an interpretation when applied to ancient sedimentary rocks, and may become a descriptive expression when applied to modern sedimentary facies (for example, crevasse splay facies). Of course, there is no rigid separation between facies and lithofacies, though Moore (1949) initially preferred the terms to be used in a strict sense. It remains more acceptable, however, to describe lithofacies and biofacies separately.

A sedimentary sequence of a particular stratigraphic unit usually grades laterally into some different sedimentary facies. These facies changes are divided into such lower levels of facies as subfacies, being expressed as Facies A, Facies B, Subfacies a, Subfacies b, and so on. Each of them reflects sedimentary environments. Therefore, it becomes possible to reconstruct original sedimentary environments more exactly by combining some lithofacies. For example, a vertical association of the sandstone facies, the siltstone facies, and the claystone facies indicates the typical fining-upward sequence that may suggest fluvial and transgressive shelf environments. Such a facies association is a useful and effective help for defining environments.

A pioneer work on facies analysis in order to reconstruct sedimentary environments is represented by J. F. M. de Raaf *et al.* (1965). In it they distinguished eleven facies in the lower Westphalian (Carboniferous) Abbotsham Formation in North Devon, England, on the basis of lithological, structural, and organic aspects. It became clear that the eleven facies recur in repetitive patterns, which suggested that the environment of deposition of these cycles ranged from a moderately deep basin, through an advancing coastline, to coastal plains.

## 5.2 Walther's Law

It is important in the concept of facies to recognize that mutually different sedimentary facies are the products of different sedimentary environments. The seashore environment in a certain area, for example, changes its position with time, and its facies boundary, as indicated by sediments, will also migrate. As a result, certain sedimentary facies extend to cover underlying strata, which were deposited in different environments. Here is a most fundamental principle in stratigraphy: that changes of environments are exhibited both by lateral changes of sedimentary facies and in vertical successions of facies. It was Johannes Walther (1860–1937) (Fig. 5.1), a German geologist and palaeontologist, who first described this concept. He argued that when in a certain area sediments of a particular sedimentary facies are succeeded by sediments of a different facies in a geological cross-section, then each of them was deposited in what had been horizontally different but contemporaneous environments (Fig. 5.2). Middleton (1973) summarized this concept as follows: 'The various deposits of the same facies areas and similarly the sum of the rocks of different facies areas are formed beside each other in space, though in cross-section we see them lying on top of each other.' This is the concept well known as Walther's Law of the Correlation of Facies, or briefly, Walther's Law. It is noteworthy that the Law is often

**Fig. 5.1** Johannes Walther (by courtesy of Prof. Gerhard Einsele, Universität Tübingen).

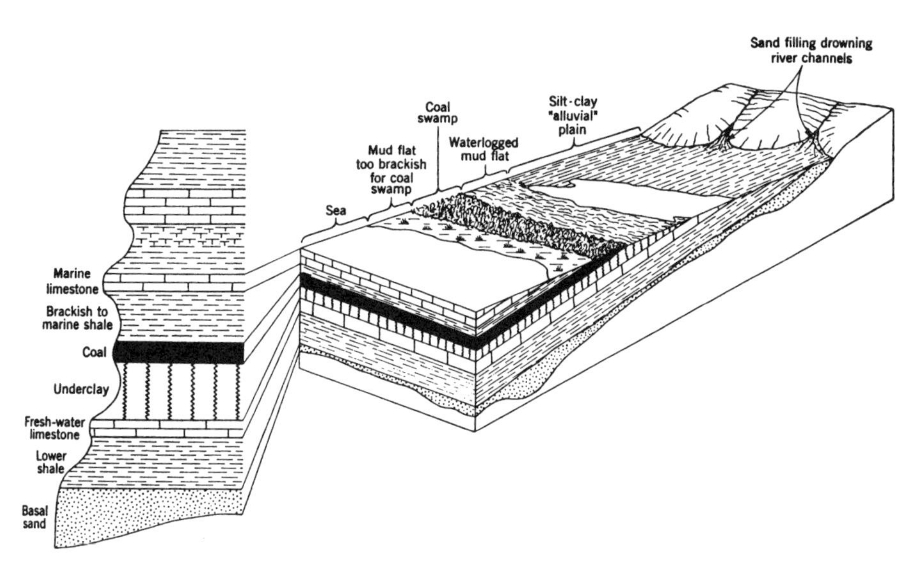


**Fig. 5.2** Simple presentation of Walther's Law, implying that vertical sequences of facies were originally formed in horizontally different environments (after Blatt *et al.*, 1972).

summarized even more simply as 'The vertical succession of facies duplicates the lateral distribution'.

Walther preferred to use the term 'ontology' for the study of present-day natural phenomena (Middleton, 1973). This term is still used in philosophy as the meaning of existence. He explained that, because the present is the key to the past, the ontological method is important in the study of modern environments and processes, and contrasted the term ontology with palaeontology. In this respect, Walther, who recognized the importance of comparative sedimentary facies in geology, deserves high appreciation. Walther's methods of the study of sediments and sedimentary rocks were to be developed further in Russia (*see* Chapter 6.1.2).

## 5.3 Facies analysis and facies models

Depostional environments are characterized by a particular geomorphic setting which is controlled by a particular set of physical, chemical and biological processes during sedimentation. The physical or dynamic process is expressed by basin geometry, depositional materials, water depth and temperature. The chemical process includes pH, Eh, carbon dioxide and oxygen in water. The biological process is the sum of the organic activities. These sedimentary facies are also defined as lithofacies and biofacies. They are well exhibited by vertical facies changes such as the fining upward and coarsening upward sequences (Potter, 1967; Hallam, 1981; Reading, 1986; Walker and James, 1992). The fining upward sequence is represented by fluvial deposits. As rivers are most important for transporting clastic detritus, the analysis of their sediments and fluvial environments is very important (Bridge, 1993).

Characteristics of lithofacies reflect physical and chemical processes at the time of deposition of sediments, and biofacies record biological activities during and immediately after deposition. It is, therefore, possible to reconstruct the environments at the time of deposition through the analyses of physical, chemical and biological aspects of sediments and sedimentary rocks (Hallam, 1981). This work is called facies analysis. It is generally said that facies analysis is the work of interpreting sedimentary processes and sedimentary environments by correlating facies distinctions with facies models.

The description of sediments and sedimentary rocks in the field should record their every aspect in as much detail as possible. For the purpose of describing lithofacies, the works of Potter and Pettijohn (1963) and Fritz and Moore (1988) are helpful regarding grain-size distribution and sedimentary structures. Bromley (1990) has done the same for biofacies.

The facies distinction follows the facies description. The facies distinction should be made objectively. For deciding the boundary between different facies, such features as erosional surfaces with channel structures, upper and lower surfaces of intercalated non-clastic layers, abrupt changes in grain size, and differences in internal sedimentary structures in spite of same grain size, may be useful criteria.

Why is the facies model the last stage of facies analysis? Walker (1979) defines the facies model as a synthetic demonstration of a particular sedimentary environment. It is a standard description of the products of that sort of environment; it is predictable, and it can become a standard even for the future.

The facies model can construct standards at various levels. For example, in sandstone successions, such levels may be identified as the sandstone facies model in channel structures, channel sandstone model in sand bars, various forms of sand bars depending on variable river beds, various river beds due to regime conditions, flow models of rivers under the combined influences of climate, topography and geology, and so on. Thus, the scale of models becomes important at the final stage of environmental reconstruction. The model scale shown below may be helpful for environmental reconstruction:

Complex environment: (e.g. the slope – submarine fan – abyssal plain model);

Single environment: (the submarine fan model);

Minor environment: (the mid-fan lobe model);

Complex sediments: (the turbidite facies model);

Sedimentation unit: (the turbidite unit-bed model).

For interpreting sedimentary characteristics and their environments, vertical facies sequences are important. Two fundamental types of vertical facies, for example, are the coarsening-upward successions and the fining-upward successions, which are the most important basic features of ancient depositional environments (for example, Potter, 1967; Bridge, 1993).

As depositional environments, deltas are very important. It is mainly because deltas and estuaries are very sensitive to changes in sea level which result in varieties of delta environments. Coarse-grained deltas with steep foreset deposition are called Gilbert-type deltas, named after G. K. Gilbert who studied Lake Bonneville (Gilbert, 1885). Gilbert-type deltas consists of a *topset* deposit with coarse-grained fluvial or alluvial sediments, a *foreset* deposit on a prodeltaic slope and a *bottomset* deposit on the deep floor at the toe of the foreset. The bottomset deposit consists of finer material, sometimes deposited by turbidity current. Barrell (1917) followed Gilbert and gave a classic account of marine deltas. A similar structural relationship of delta sedimentation was recognized by J. L. Rich, who characterized the megafacies of delta into *undathem* (neritic), *clinothem* (slopes) and *fondothem* (bottom) facies. The corresponding structural terms were *undaform, clinoform* and *fondoform*. L. M. J. U. van Straaten (1961, 1964) and E. D. McKee (1979) also introduced new trends in recording the details of facies. The term *diastem* was proposed to show the time break represented by submarine nondeposition (Heckel, 1972).

For the description of lithofacies, the method of lithofacies code proposed by Andrew D. Miall (1978) (Fig. 5.3) is very practical and useful. This method is to give a notation to a lithofacies characterized by particular sedimentary structures, and to correlate it to a modelled sedimentary environment. Miall (1978) proposed nineteen codes for fluvial sediments. For example, the stratified gravel with planar cross beds is given a code Gp, which is a combination of G from gravel and p from planar, and then, its genetic environment is correlated with small-scale,

**Fig. 5.3** Andrew D. Miall (photo by H. Okada).

**Fig. 5.4** H. N. Fisk (Russel, 1965). (*Reproduced by permission of the Geological Society of America.*)

channel-fill sediments. How to give codes to lithofacies is explained in detail by Yagishita (2001). Similarly, J. L. Wilson (1975) proposed twenty-four codes for the description of microfacies of carbonate sediments.

It can be said that a typical study of the facies model was initiated for the modern Mississippi Delta. A detailed framework of the modern deltaic environments of the Mississippi Delta presented by H. N. Fisk (1908–1964) (Fig. 5.4) is still a benchmark for the study of deltaic sediments (Fisk *et al.*, 1954). Since then, the recognition of the correlation between the present and the past has become essential to the study of sedimentary geology. Figures 5.5 and 5.6 show examples of the environmental analyses of the bird-foot delta by Fisk *et al.* (1954).

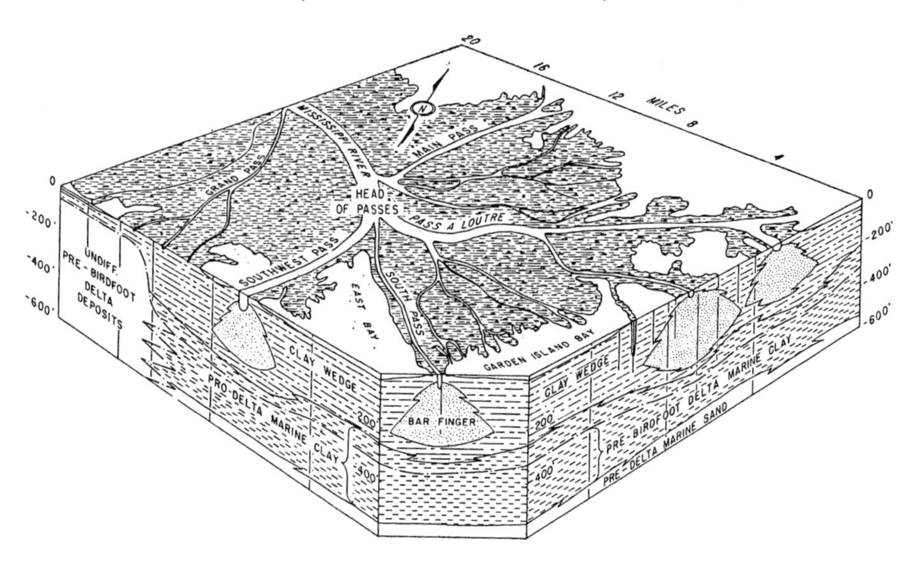


**Fig. 5.5** Depositional framework of the Mississippi Delta (after Fisk *et al.*, 1954). The depths are indicated in feet.

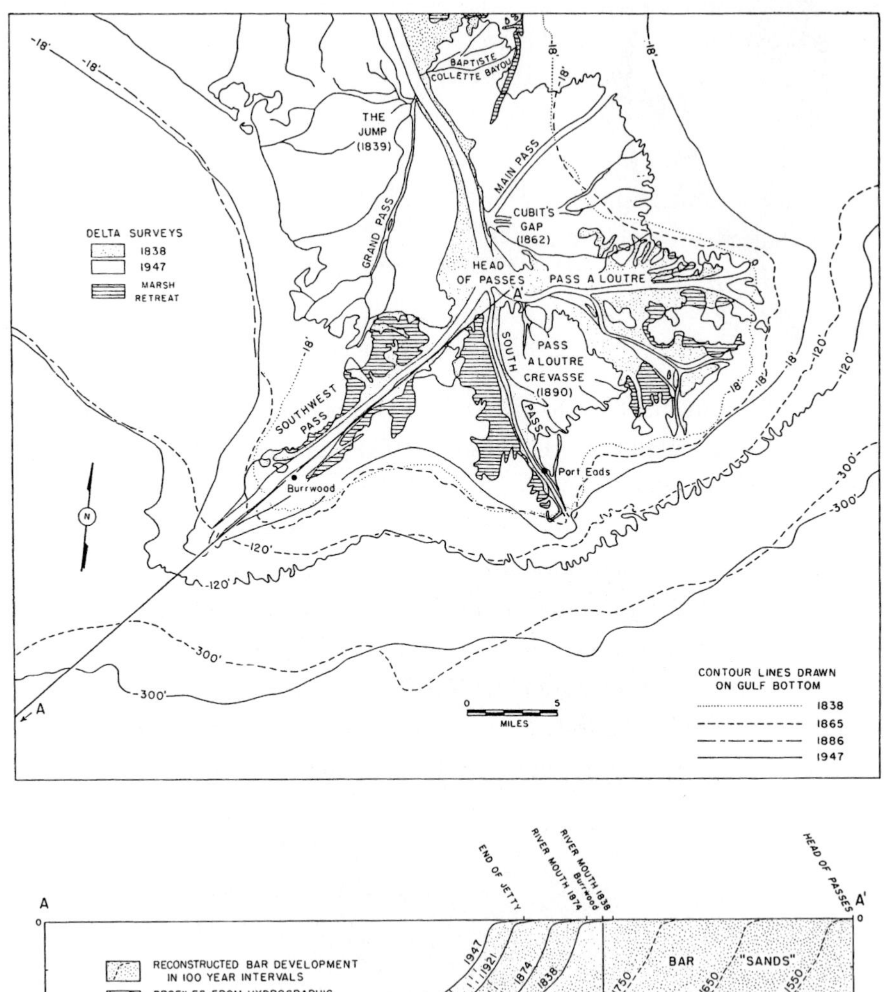


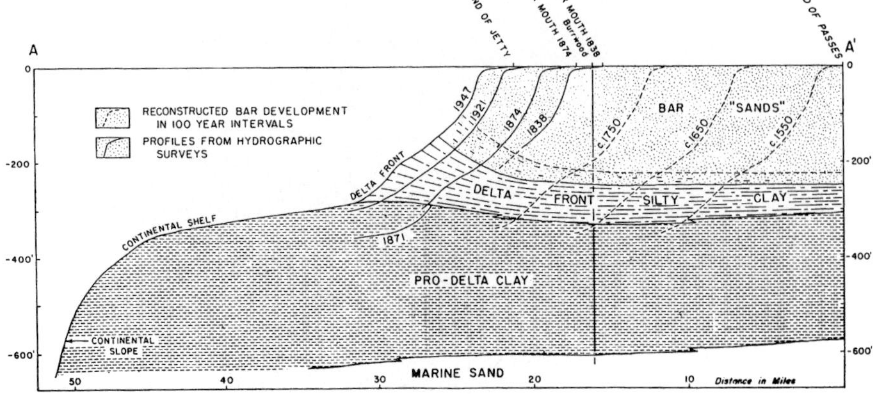


**Fig. 5.6** Progradational pattern of the birdfoot delta of the Mississippi River (after Fisk *et al.*, 1954). The depths are in feet.

Other such studies include the analyses of the flow systems of the Klarälven River in Sweden by Å. Sundborg (1956), the study of the intertidal zone of the North Sea by L. M. J. U. van Straaten (1954), the reconstruction of the Lower Cretaceous Wealden facies by P. Allen (1959), the analyses of the fluvial environments of the Old Red Sandstone by J. R. L. Allen (1964) (Fig. 5.7) and his study of the modern Niger Delta (Allen, 1965b), deltaic systems by W. Nemec and R. J. Steel (1988), the flysch facies model by Dzulynski and Smith (1964), the submarine fan depositional model by E. Mutti and R. Ricci-Lucchi (1972), and the tidal facies model by G. deV. Klein (1977). All represent benchmark studies of facies models. Figures 5.8 and 5.9 show a good example of the sedimentary facies analysis of the typical arcuate delta, the Niger Delta, by J. R. L. Allen (1965b, c), a professor of the University of Reading in England.

Furthermore, Reading (1986) and Walker and James (1992) proposed facies models corresponding to sea-level changes, based on the sequence stratigraphy concept in a variety of sedimentary environments.

**Fig. 5.7** John R. L. Allen (by courtesy of Sedimentology Research Laboratory, Reading University, 1965).

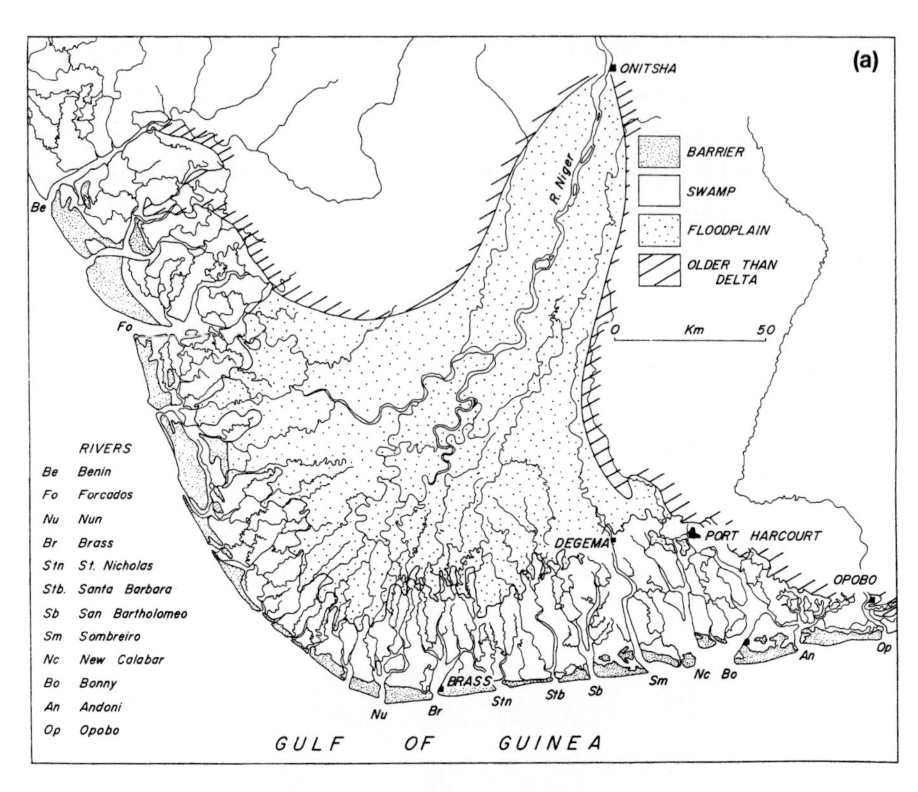


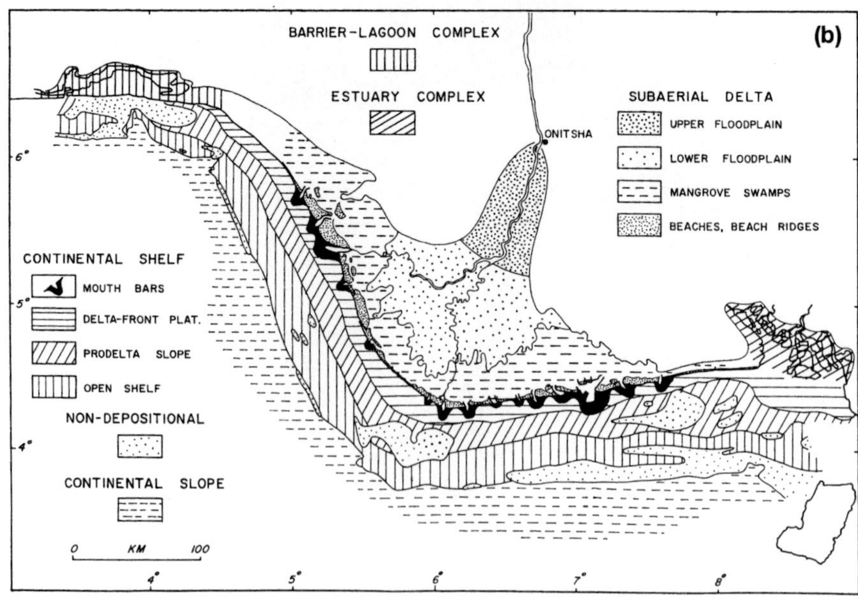


**Fig. 5.8** (a) Morphology of the Niger Delta (J.R.L. Allen, 1965c), a typical example of the arcuate delta. (*Reproduced by permission of the A.A.P.G.*) (b) Depositional environments of the Niger Delta (J. R. L. Allen, 1965b). (*Reproduced by permission of Netherlands Institute of Applied Science.*)

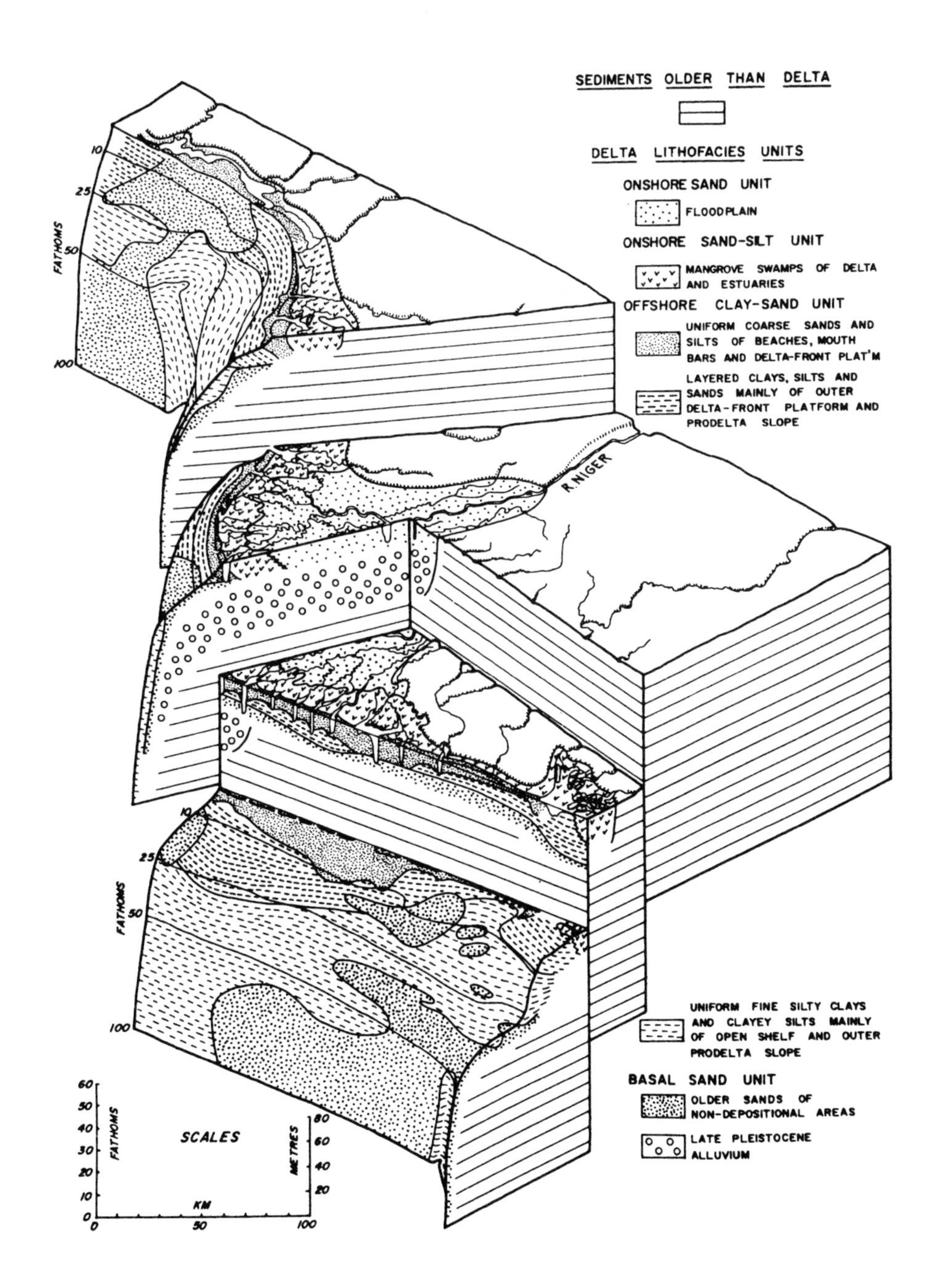


**Fig. 5.9** Block diagrams showing the mutual relationship between the delta lithofacies units of the Niger Delta (J. R. L. Allen, 1965b). (*Reproduced by permission of the A.A.P.G.*)

## 5.4 Lithology

The term *lithology*, from the Greek *lithos* (= stone), refers to rocks in general and indicate the general descriptive term of a hand specimen as well as of an outcrop. Lithology is in general defined as a discipline for studying the characteristics of sediments and sedimentary rocks and their genetic environments on the basis of lithofacies. This discipline achieved a great deal of development in Russia in the early twentieth century; this is discussed in the next chapter. Incidentally, the English word 'lithology' can refer to all kinds of rocks comprising the earth's lithosphere. According to G. V. Middleton (1978; personal communicaton, 20 June 2002), the term implies two meanings. The first refers to the petrology mainly based on the description of samples and generally used for the description of sedimentary rocks (Bates and Jackson, 1987). In particular, the term is applied to the description of the characteristics of sedimentary rocks as specimens, structures of exposed rocks and features other than contained fossils. This is the most common usage of the term. The second meaning relates to the description of lithofacies, as Walther intended (Middleton, 1973). In other words, it was used for describing various features of sedimentary rocks controlled by sedimentary environments. Such a usage of the term has become common only since the 1970s. Therefore, it may be said that the term *lithology* had originally excluded the implication of sedimentary environments in both Britain and the United States.

In Japan, *Chisogaku* (a Japanese translation of 'lithology') had been used in reference to the comprehensive nature of sediments and sedimentary rocks including the composition of sediments, lithofacies, sedimentary structures, and depositional processes. Hanjiro Imai clearly indicated the meaning of the term in his book *Chisogaku* published in 1931, defining *Chisogaku* as the study of the entire nature of sediments and sedimentary rocks. This concept is akin to the broad implication of sedimentation as used by W. H. Twenhofel in his books *Treatise of Sedimentation* (Twenhofel, 1932) and *Principles of Sedimentation* (1950) and to Hakon Wadell's concept of 'sedimentology' (Wadell, 1932b). Masao Minato (1953) also published a book with the same title *Chisogaku,* in which he defines the term in a more dynamic sense as embracing every feature of strata with respect to lithofacies, biofacies, ecology, structure, chronology and tectonic development.

In conclusion, it can be said that the concept of lithology as widely used in Russia and Japan had the same meaning as 'sedimentology'.

# 6

# Establishment of sedimentology

## 6.1 Genealogy of sedimentology in Europe, Russia and America

The study of sediments and sedimentary rocks before the acceptance of the term *Sedimentology* can be associated with three geographically distinct groups of researchers: in Western Europe, Russia and America. Each regional group seems to have maintained control of its own research methodology. The academic features of these groups are expressed by the European school, Russian school and American school, respectively. Characteristics of each of these schools and their backgrounds are described below.

### 6.1.1 European school

The early study of sediments and sedimentary rocks in Western Europe is characterized and represented by microscopic sedimentary petrography, which originated and became established in Britain. As mentioned before, it was Sorby who introduced the method of polarized microscopic observation of thin sections of rocks into geology and applied it successfully to a systematic study of Jurassic calcareous sandstones in Yorkshire, England. Based on his achievements, microscopic sedimentary petrography slowly became popular in Britain, though it took about fifty years for it to become widely used. The British were so focused on palaeontology and biostratigraphy that few people paid much attention to other aspects of sedimentary rocks! However by 1900–1920, sedimentary petrography flourished accompanied by the recognition of the value of heavy minerals. Thus, Britain gained fame as the centre for the study of microscopic sedimentary petrography and gradually this methodology spread into Western Europe and flourished there.

During the period from the 1920s to the 1940s, many important textbooks on sedimentary petrography were published in Europe: for example, in Britain by Hatch and Rastall (1913), Milner (1922, 1929), Boswell (1933); in France by Cayeux (1929a, b); and in Germany by Correns (1939, 1949). Among these publications, Lucien Cayeux's books on the petrography of sedimentary rocks in France, which include many fine microscopy figures of various sedimentary rocks, became masterpieces.

Such a trend of research in sediments and sedimentary rocks in Western Europe without doubt characterizes the 'European school'. The founder of microscopic sedimentary petrography in Western Europe was without doubt Prof. Henry Clifton Sorby, Professor of Sheffield University, England. From his first petrographic study of a calcareous grit in the Yorkshire coast, he went on to describe microscopic features of calcareous grits of various ages, as well as oolites and sandstones. For his initiation of microscopic petrography he received the honour of the Wollaston Medal from the Geological Society of London in 1869. Further, in 1907 when the Geological Society of London was celebrating its centenary, Sorby was elected President of the Society and in his Presidential Address he commented on his life's work (Sorby, 1908). On the occasion of his Address many eminent geologists, including Archibald Geikie and Ferdinand Zirkel, paid formal tribute to him and declared Sorby to be the 'Founder of Petrography'.

Since then, Sorby has also been described as the 'Father of Petrography' by Folk, 1965. Sorby's great scientific powers were also directed to sedimentation and transportation processes, and included the application of experimental methods. Sorby's observations of current structures and sedimentation led the study of sediments and sedimentary rocks into a dynamic field of geological sciences (Sorby, 1908; Wilcockson, 1945). Wilcockson (1945), Summerson (1976), Allen (1993) and Okada (2004) have all praised Sorby as the 'Father of Sedimentology'. M. R. Leeder (1998) has, however, disputed this accolade, seeking instead to recognize the significance of Lyell's contribution to sedimentology by referring to Volume 1 of Lyell's *Principles of Geology* (1830–1833) which included field sketches of cross-stratification (*see* Fig. 3.1. No. 34). The present author does not accept Leeder's views and asserts that the title 'Father of Sedimentology' must belong to Sorby, and would add that Sorby added to our knowledge of plants, marine life, other aspects of natural history and perhaps (*see* Simons, 1952) was the founder of metallurgy. It was indeed proper that the Geological Society of London awarded Sorby the Wollaston Medal in 1869 and made him its President in its Centenary Year – few scientists can have been held in such veneration.

### 6.1.2 Russian school

During the period when the European school was at its zenith, Russian geoscientists were developing their own method of study of sediments and sedimentary rocks, one which differed from that of Western Europe and that of America. The Russian territory was vast, and mostly in the cold climate zone. In order to raise agricultural productivity, they had to understand regional soil characteristics and the lithologic features of basement rocks. In was in such a circumstance that Johannes Walther (1893, 1894) exerted a decisive influence upon the research approach to sediments and sedimentary rocks in Russia (Middleton, 2003). According to Middleton (2003), the influence of the *Challenger* reports (*see* Chapter 7.2) was also strong.

Middleton (2003a) says Y. V. Semoilov in 1923 wrote of lithology as a systematic science. He also says that the 1920s and 1930s were the main period of the acceptance of sedimentology in Russia. The first textbooks on sediments in Russia were published in 1932 and 1934.

Research methods attaching great importance to the facies analyses of sediments and sedimentary rocks grew rapidly in Russia. In consequence, all studies of sediments and sedimentary rocks came to be called 'lithology'. The periodical journal *Lithology and Mineral Resources*, which started in 1963, was a symbolic result of such Russian research trends. This approach to research in sediments and sedimentary rocks should be called 'the Russian school'. The following is an interesting definition of lithology proposed by N. M. Strakhov (1970): 'Lithology is the interpretation of sedimentary rocks in terms of facies and environments of deposition and to develop larger theories of sedimentary rocks.' It was Nikolai Mikhailovich Strakhov (1900–1979) (Fig. 6.1), an Academician of the Russian Academy of Sciences, who emphasized soil formation and observed chemical and biological actions in relation to lithofacies and climate zones. He recognized the genesis of sedimentary rocks in terms of comparative lithofacies generated by climatic influences (Strakhov, 1962, 1970).

Incidentally, at that time, in the late 1970s, the term *sedimentologia* first began to be used in Russia (Kazanskii, 1976).

**Fig. 6.1** N. M. Strakhov (*reproduced by permission of Prof. Galina L. Kirillova, Institute of Tectonics and Geophysics, Far East Branch of Russian Academy of Sciences, Khabarovsk*).

### 6.1.3 American school

Studies of sedimentary rocks had become active in America in prospecting for oil and natural gas as the motive energy for the New Industrial Revolution early in the twentieth century. Studies of textures and granulometry of sediments were related to the migration and entrapment of hydrocarbons, thus grain size, porosity, and particle shape were of particular interest, as was sedimentary mineralogy, especially of heavy minerals.

For example, classification of the grain sizes of clastics was based on Udden-Wentworth grade-scale (J. A. Udden, 1914; C. K. Wentworth, 1922). It was modified to the present Wentworth grain-size scale by Wentworth (1922) (Fig. 6.2), in which he adopted the power of 2 in millimetres in diameter. This grain-size scale, however, created difficulties when handling such a wide range of grain size from a boulder more than 256 mm in diameter to clay less than 1/256

**Fig. 6.2** Chester Keeler Wentworth
(*reproduced by permission of the Geological Society of America*).

mm in diameter. In order to alleviate this problem, W. C. Krumbein (1934) proposed a new method of logarithmic transformation of the Wentworth grain-size scale, called the phi scale, which is obtained as follows: $ø = -\log_2 d$, where $d$ = diameter of the grain in millimetres. This easy presentation of grain-size data made it possible to develop the quantitative study of the granulometry of sediments. After the establishment of the grain-size scale remarkable progress in the study of grain-size analysis or granulometric analysis of clastic sediments was made in America, followed by the development of statistical data-analysis and presentational methods. Among them, D. L. Inman (1952), R. L. Folk and W. C. Ward (1957) and Folk (1965) contributed much to the establishment of such statistical parameters of grain-size distribution as mean, standard deviation, skewness and kurtosis.

In addition, studies of grain shape were developed: quantification of roundness and sphericity by H. Wadell (1932a, 1933b); the silhouette chart for estimating roundness of gravels by W. C. Krumbein (1941a); and the visual comparison chart of roundness and sphericity of sand grains by M. C. Powers (1953). Furthermore, the electron microscope was applied to studies of the surface texture of grains becoming a unique research method for analysing depositional processes and recognizing sedimentary environments (Krinsley, 1962; Porter, 1962; Bull, 1986).

There was great interest in the 1920s and 1930s in the possibility of correlation of sandstones using accessory minerals. This line of research was also carried out in the petroleum industry. Sedimentary mineralogy competed with micropalaeontology as to which technique would be best.

Studies of clastic particles and the mutual relationship between grains made great progress in the United States. A good synthesis of these researches was presented by Trask (1939), who edited case studies on the relationship between size distributions and transportation in marine sediments. The research methods, as developed in the United States concerning grain morphology and sedimentary texture, can be called 'the American school'.

The development of sedimentary research, closely related to oil prospecting, progressed rapidly, and led to the creation of the Society of Economic Paleontologists and Mineralogists (SEPM) in 1926 as a junior organization of the American Association of Petroleum Geologists. The Society had, as its periodical publications, the *Journal of Paleontology* and the *Journal of Sedimentary Petrology (JSP)*, which started in 1927 and 1931 respectively.

It is noteworthy that a major change within the SEPM took place at its Council meeting held at Keystone, Colorado in 1988. Although those present acknowledged the important role the Society had played hitherto in the development of geology, it was proposed that the Society of Economic Paleontologists and Mineralogists should be renamed as the Society for Sedimentary Geology in order to better reflect its interests. This proposal was approved at the general meeting of the SEPM and its new name became 'SEPM (Society for Sedimentary Geology)'. At the same time, the name of its regular publication *Journal of Sedimentary Petrology* was changed to *The Journal of Sedimentary Research*.

## 6.2 Tripartite study of sediments

When reviewing the development of sedimentology, the year of 1950 seems to have been pivotal: a time when there were major changes in the methods of study of sediments and sedimentary rocks. About this time the subject moved away from the influence of three distinct schools to the acceptance of three themes of research.

Before 1950, three major schools of sedimentary research, characterized by the respective regions of Western Europe, where sedimentary petrography was developed (European school); of Russia, where lithology was a predominant research method (Russian school); and of the United States, where studies of sedimentary textures and sedimentary processes were prevalent (American school). However, responding to the rise of sedimentology after 1950, these research methods have been blended together leading to invigorating sedimentary researches and guaranteeing the rapid progress of sedimentology as a fundamental discipline of the geological sciences.

When looked at from a different viewpoint, trends of sedimentary researches after 1950 are characterized by (i) direct observations of modern and ancient sediments and their environments (lithofacies analysis) and their modelling (field sedimentology), (ii) experimental modelling of sedimentary processes under various conditions (experimental sedimentology), and (iii) mathematical modelling and computer simulation (mathematical or theoretical sedimentology). Such trends show sedimentology to be innovative and apposite.

The study of modern sedimentary environments, such as studies by H. N. Fisk *et al.* (1954) of sedimentary environments of the Mississippi Delta, environmental analyses of the Niger Delta by J. R. L. Allen (1965b), studies by L. V. Illing (1954) of modern carbonate deposits in the Bahama Bank, are all good examples of the major achievements. Influenced by these studies, many sedimentary researchers since then have emphasized the comparison between the present and the past as their main themes, as exemplified by modern and ancient geosynclinal sedimentation (Dott and Shaver, 1974), modern and ancient alluvial sedimentation (Nilsen, 1985), modern and ancient tidal sedimentation (Flemming and Bartholomä, 1995).

As to experimental sedimentology, studies of turbidity currents by the experiments of Ph. H. Kuenen in 1950 were epoch-making and demonstrated the importance of experiments in geology (Kuenen and Migliorini, 1950). Later Kuenen carried out experimental studies of sedimentary structures such as sand volcanoes, load casts, pseudo-nodules and others (Kuenen, 1958a), and, further, drew attention to the importance of experiments in sedimentology such as the abrasion features of clastic particles under various conditions (for example, Kuenen, 1956, 1965). Since then, such *in situ* observations on the movement of particles under natural conditions and the field observations of grain abrasion (for example, Carr, 1971) as well as interactions observed in flume experiments between the grain size of clastics and flow regimes, observations of bed-forms of rivers, and examinations of the constructional mechanisms of particular sedimentary structures have developed remarkably. At the same time, the scale problem in such studies has been examined in detail (Strahler, 1958; Williams, 1970).

Theoretical studies in sedimentology have also made remarkable progress in the statistical treatment of sedimentological data and such stochastic or random process models as bed-thickness distribution analysis and Markov chain analysis. A good review in these aspects was made by S. Mizutani (2000). In such mathematical and theoretical studies, both the time factor and the scale factor have to be taken into account.

It may be concluded that, based on the scientific organization of field, experimental and theoretical studies, a modern sedimentology has been established.

## 6.3 Controversy about the term 'Sedimentology'

Before the achievement of the establishment of modern sedimentology, a curious detour was necessary. It was the struggle over the acceptance of the term 'sedimentology'.

The application of the term 'sedimentology', which is expressed as *Taisekigaku* in Japanese, was first proposed by Hakon Wadell (1895–1962) (Fig. 6.3), who was a graduate student of Professor F. J. Pettijohn at the University of Chicago at that time, as the word to express sedimentary researches in all aspects of sediments and sedimentary rocks (Wadell, 1932b), in place of 'sedimentation' and 'sedimentary petrology', both of which had been commonly used to describe sedimentary studies up to that time. However, the term 'sedimentology' was not coined by him, but, rather as Wadell (1933a) said: 'the term *sedimentology* was first introduced some ten years ago by A. C. Trowbridge, who, in *American Men of Science*, has listed sedimentology as one of his major subjects.' According to Pettijohn, Trowbridge, of the University of Iowa, had at that time suggested *Sedimentology* as the title of the new journal instead of *Journal of Sedimentary Petrology*, which was launched in 1931 by the Society of Economic Paleontologists and Mineralogists (*see* section 6.1.3 above: the American school).

**Fig. 6.3** Hakon Wadell (Bretz, 1963).
(*Reproduced by permission of the Geological Society of America.*)

Strong opposition, however, was evoked against the new term 'sedimentology' in the Committee on Sedimentation of the U.S. National Research Council. The Committee prohibited the usage of the term mainly for the reason that it contains roots from two languages; that is, a derivation from both Latin and Greek. The chairman of the Committee was Professor W. H. Twenhofel of the University of Wisconsin (1875–1957) (Fig. 6.4), who retained a strong influence in the study of sedimentation and sedimentary rocks in the United States. The strong objection came principally from Twenhofel (Pettijohn, 1984). As a consequence, the term 'sedimentology' was not accepted in America (Pettijohn, 1984).

The dispute was also expanded by M. I. Goldman (1950) and J. L. Hough (1950) in America. Their major objection was that the term 'sedimentary petrology' was already firmly established, and that the term 'sedimentology' was an ugly hybrid and inappropriate word tying together both Latin and Greek (Goldman, 1950). Wadell (1933a) differed strongly showing that the word 'terminology' was itself a compound word of Latin and Greek, and thus concluded that there was no reason for disqualifying the term 'sedimentology'. He further asserted that 'sedimentology' was superior to 'sedimentation' as the term for the subject,

**Fig. 6.4** William H. Twenhofel (courtesy Prof. Robert H. Dott, Jr., University of Wisconsin). (*Reproduced by permission of the A.A.P.G.*)

and that hybrid word formations are frequently used in science. (*Author's note:* It is ironic that the term *petroleum* which had been in wide use – for instance the American Association of *Petroleum* Geologists – also took its roots from both Latin and Greek!)

While the term 'sedimentology' was in serious dispute in the United States the organization of the subject had been progressing steadily in Western Europe, mainly under the leadership of Dutch and British sedimentologists. It was Professor D. J. Doeglas of the Agricultural University of Wageningen in the Netherlands who led the organization of sedimentology in Europe. Doeglas (1951a) criticized the movement to reject the term sedimentology in the United States. He said: 'From a terminological standpoint there seems to be no great objection to the term "sedimentology". Spectrology, volcanology, radiology, criminology and many other words have a Latin–Greek combination. The second objection of Goldman, however, is more serious. It might be directed against our standing as petrologists. What is there in favor of its use? In my opinion it is the historical evolution of our science. In the old days the term "sedimentary petrology" was not used. The term "mineralogy" of sedimentary rocks was in use. Around 1920 the term "sedimentary petrology" became found more commonly in literature.'

Dutch geologists were leading in sedimentology in the 1950s, contributing to the establishment of the International Association of Sedimentologists. Why could Dutch geologists be a driving force for sedimentology? H. G. Reading (1974) answered this question as follows: 'It was not only because the surface of Holland is largely Holocene and most Dutch geologists are brought up to think of the Pleistocene as "basement"; the Dutch were also taught to take ancient sediments as seriously as modern sediments. Kuenen in his *Marine Geology* (Kuenen, 1950b) has a long section on geosynclines, mentions Carboniferous cycles, and the relevance of his book to ancient rocks is obvious. The American, Shepard, on the other hand, never mentions an ancient rock sequence in *Submarine Geology* (Shepard, 1967) and at no point is there any suggestion that his book might have a bearing on investigations of ancient rocks.'

Unfortunately, the rejection of the term 'sedimentology' in America continued up to the 1970s. As mentioned before, W. H. Twenhofel (1941) had stubbornly adhered to 'sedimentation', 'sedimentary mineralogy' and 'sedimentary petrology'. Under such circumstances, Krumbein and Pettijohn (1938) and Krumbein and Sloss (1953) had tried to use the term 'sedimentology' in their publications, but their attempts had never succeeded. Pettijohn (1975, p. 2), however, carried a chapter on 'History of Sedimentology' in his revised edition of *Sedimentary Rocks*.

## 6.4 Organization of sedimentology

Sedimentary researchers in Western Europe began to promote the organization of the new discipline of 'sedimentology' soon after World War II. On the occasion of the 18th International Geological Congress held in London in 1948, British sedimentary petrologists headed by Percival Allen (Fig. 6.5), a professor of the University of Reading, U.K., arranged a meeting of the sedimentary petrologists who had participated in the Congress. There, Allen proposed two important issues for discussion: (1) a sort of international union of sedimentary petrologists and (2) the initiation of regular international meetings between them. This proposal had a positive response from the attendees who agreed that the details of the

**Fig. 6.5** Percival Allen (courtesy Sedimentology Research Laboratory, University of Reading, 1965).

organization should be decided at the next International Geological Congress. An International Secretariat for Sedimentary Petrology was, however, created, and Professor D. J. Doeglas (1906–1974) (Fig. 6.6) was elected as the International Secretary General. Thus, preparation for establishing an international union of sedimentologists had started smoothly. The details of this process are documented by D. J. Doeglas (1976). As an advocator of this international organization, Allen's contribution was significant. He was not only a strong supporter of sedimentology, but also the founder in 1961 of the Sedimentology Research Laboratory at the University of Reading (presently the Postgraduate Research Institute of Sedimentology, University of Reading) (Fig. 6.7) in England, which was the first institute of sedimentology in the world and since has established a key position in world sedimentology.

Against this background, as early as 1946, a First International Congress of Sedimentology had already been held in Ghent, Belgium, which had 'The Geology of Recent Deposits in Western Europe' as its main theme. After that,

**Fig. 6.6** D. J. Doeglas (Nota, 1974).

**Fig. 6.7** Sedimentology Research Laboratory, University of Reading in 1965 (presently, the Postgraduate Research Institute for Sedimentology, University of Reading) (courtesy Sedimentology Research Laboratory, University of Reading).

two meetings were held before the creation of the International Association of Sedimentologists; the Second Congress hosted by French sedimentologists and sedimentary petrologists at La Rochelle in France in 1949 on the topic of *Sédimentation et Quaternaire* and the Third Congress was at Groningen and Wageningen in the Netherlands in 1951, on the theme of 'Recent Sediments'. These meetings were small and were limited to participants from Western European countries.

Promoters of sedimentology in Europe considered the term 'sedimentary petrology', to which American researchers had adhered, to be too restrictive, whereas the term 'sedimentology' included both the field study and laboratory study of sediments, ancient and modern. D. J. Doeglas (1951a) considered that an early focus on the term 'sedimentology' was essential in order to develop a broad range of sedimentary researches. He argued that sedimentology, but not sedimentary petrology, indicates a study of all the properties of sediments and their variations in a horizontal and vertical sense (Doeglas, 1951a). He further admitted that even at the time of the Third International Congress of Sedimentology, he was still wondering whether the word 'sedimentology' would be adopted or not.

As had been expected, however, an international union of sedimentologists named the International Association of Sedimentology was founded with about 300 members at the 19th International Geological Congress at Algiers in 1952 (Doeglas, 1976). The first President of the Association was D. J. Doeglas. However, he resigned the Presidency before the 4th International Sedimentological Congress (ISC) in Göttingen in 1954, being replaced by C.W. Correns (Germany) for the period 1954–1958. The 5th ISC was held in Switzerland. F. P. Shepard (U.S.A) was elected as the President for 1958–1962. Soon, in 1963, the name of the Association was changed to 'the International Association of Sedimentologists' (still IAS) instead of '. . . Sedimentology'. After the 4th ISC in 1954, it was agreed that the congresses should take place every four years and not be in the same year as the International Geological Congress. The inauguration of the Association coincided with the start of its own journal *Sedimentology* in 1962. Published bi-monthly it has become a focal publication for sedimentologists worldwide. Subsequently in 1988, in collaboration with the European Association of Exploration Geophysicists (now the European Association of Geoscientists & Engineers, IAGE), the International Association of Sedimentologists started publication of a quarterly journal entitled *Basin Research*. This journal addresses the problems of sedimentary basins as geodynamic entities.

The IAS was and remains strongly European in character due, in part, to the controversy over the term 'sedimentology' and the subsequent creation by Europeans of the International Association of Sedimentologists. This circumstance seems to have influenced the reorganization and internationalisation of the SEPM of the United States. The European Regional Meeting held at Davos in Switzerland in 2001 was the twenty-first meeting. The Davos meeting was attended by 467 sedimentologists from 52 countries, thus demonstrating a trend of globalization of the regional meetings (Bosence, 2001). The organization of such regional meetings was encouraged and positively supported by the International Association of Sedimentologists at the 12th ISC in Canberra, Australia, in 1986, but has rarely been put into operation outside Europe.

## 6.5 Establishment of sedimentology

As stated above, it is no exaggeration to say that the modernization of sedimentary researches started with the establishment of the term 'sedimentology'. For this scientific environment, the contribution by European sedimentologists has been significant. Among them, in particular, the efforts of D. J. Doeglas, who was one of the promoters of the term 'sedimentology' and stressed its importance to people and society at every available opportunity (Doeglas, 1951b, 1955), were important. The strong conviction that the acceptance of *sedimentology* would make possible an international organization of sedimentary researchers has made good. The foundation of the International Association of Sedimentologists in 1952 certainly accelerated the movement for establishing the new field in earth sciences.

It was around 1960 when this new field of geological sciences became more widely recognized. At this time, the term 'sedimentology' finds its place in dictionary pages. For example, the term was recorded in *Encyclopaedia Britannica* in 1959, and in *Webster*'s dictionary in 1961. The 1961 edition of *Webster* gives the following entry for the word: 'Sedimentology: The description, classification, and interpretation of sediments' and 'Sedimentologist: One who investigates sedimentation or studies of sedimentary processes'.

# 7

# Sedimentology and the study of the ocean floor

## 7.1 Significance of ocean floor research

It can be said that ocean sediments are the relics of time. They record earth environments at every stage of the evolution of the Earth. Hence, it is quite natural that studies of deep-sea sediments have contributed much to the significance of sedimentology. In particular, increased understanding of the sedimentary processes on the modern ocean floor greatly influenced the development of sedimentary geology. As Philip H. Kuenen (1958b) put it, there is 'No geology without marine geology'. Furthermore, it is interesting to note in this connection that the term 'sedimentology' has become popular in the United States of America mainly because of the great achievements of sedimentologists involved in the deep-sea drilling programmes of the 1970s. This chapter, therefore, will explore the relationship between sedimentology and the study of the ocean floor, bearing in mind that the present ocean floor dates back only to the Jurassic or possibly late Triassic periods.

## 7.2 Achievements of the Challenger Expedition

The year 1872 saw two seminal events in sedimentary geology – the acceptance of the geosyncline as a fundamental geological concept and the commencement of the voyage of HMS *Challenger*, which was to 'explore systematically the physical, chemical and biological features of the oceans, to determine their bathymetry and examine the nature of the ocean floor's sediments'.

The historic expedition of HMS *Challenger,* a three-masted, steam screw, wooden corvette of 2,300 tons (Fig. 7.1), was organized jointly by the Royal Society of London and the British Royal Navy. She sailed from Portsmouth on 21 December 1872, and after a voyage of three and a half years and nearly 120,000 km over the seven oceans, she anchored at Spithead, England, on 24 May 1876. Captain George S. Nares with about twenty naval officers, a relatively small and dedicated crew, and Professor Charles Wyville Thomson (1830–1882), professor of natural philosophy at Edinburgh, and six scientists were on board the *Challenger*. On Wyville Thomson's sudden death during the preparation of the report on the expedition's discoveries, John Murray (1841–1914) (Fig. 7.2), one of the members of the expedition, was nominated as the new editor.

During the voyage, using many innovative instruments, routine work included sounding the ocean floor (over 8 km was the maximum recorded depth, an astounding discovery at the time), observations of temperatures and current directions at various depths, sampling of the water and the sea life at

**Fig. 7.1** HMS *Challenger* (Linklater, 1972).

**Fig.** 7.2 John Murray (Linklater, 1972).

various levels, and dredging of samples of the bottom sediments at 362 scattered stations. The tremendous achievement of the *Challenger* had raised the curtain on modern oceanography. The final *Challenger Report* with fifty volumes and 29,500 pages, established a great monument in science. In particular, the classification of deep-sea sediments and descriptions of micro-organisms, manganese nodules (*Challenger* collected large amounts of manganese nodules from the ocean floor near the Hawaiian Islands in September and October, 1875), and cosmic dusts were important, while the monograph on radiolarian assemblages by Ernst Haeckel was epoch-making (Haeckel, 1887). John Murray and A. F. Renard (1891) showed that pelagic sediments contain very little terrigenous material, but are characterized by oozes mainly composed of such micro-organisms as radiolarians, foraminifers, coccoliths and pteropods. They further made it clear that the sediments of the deep (abyssal) ocean floor are composed of two major lithofacies: calcareous oozes and red clay. The latter is typical of the deeper floors because calcareous ooze dissolves as the depth increases. Now the depth at which calcareous components are completely dissolved is called the calcium carbonate compensation depth or CCD (Bramlette, 1961).

Such new facts on deep-sea sediments strengthened the uniformitarianism hypothesis proposed by Charles Lyell, though Lyell could not have known of any deep (abyssal) ancient sediments to which the *Challenger* samples could be compared. The new findings about the deep seas were almost entirely *new* and unexpected. Hitherto it was generally assumed that the deep seas would contain sediments dating back to early Precambrian. But this proved to be false! Further, based on the *Challenger* discoveries, Émile Haug (1900) advocated his idea of the deep-sea origin of geosynclines. Clearly the *Challenger* expedition overshadowed all previous efforts some of which had made important discoveries, at studying the deep ocean (Friedman and Sanders, 1978).

At present, oceanic or pelagic sediments are classified as follows (after Arrhenius, 1963).

**Ooze:** composed of more than 30% of micro- and nano-planktons.

**Calcareous ooze:** composed of calcareous tests of micro- and nano-planktons, such as foraminifer ooze, coccolith ooze, pteropod ooze.

**Siliceous ooze:** composed of siliceous tests of micro- and nano-planktons, such as radiolarian ooze, diatom ooze.

**Pelagic clay:** red, brown, or chocolate-coloured clay composed mainly of fine-grained terrigenous substances less than 10 microns in size and less than 20% in amount, together with decomposed tests of planktons, ultra fine particles of volcanic origin, both submarine and terrestrial in origin, desert and extra-terrestrial dusts, durable items such as shark's teeth, all often associated with manganese nodules.

**Hemipelagic sediment:** intermediate or 'half-pelagic' sediment between terrigenous sediment and pelagic clay. This is a deep-sea sediment in which more than 25% of the fraction coarser than 5 microns is of terrigenous, volcanic and/or neritic origin. It is found near the continental margin and its adjacent abyssal plain.

## 7.3 Deep-sea drilling and sedimentology

The original motivation for the Deep Sea Drilling Project as proposed by Maurice Ewing at the Lamont-Doherty Geological Observatory, Columbia University, was not to verify sea floor spreading, rather it was simply to gather data about the deep sea floor, (Horsfield and Stone, 1972). Its verification of spreading was an unpredicted outcome of DSDP. However, the Deep Sea Drilling Project (DSDP, 1968–1975; International Phase of Ocean Drilling, IPOD, 1975–1985) carried out on board D/V *Glomar Challenger* (10,000 tons, constructed in 1968) (Fig. 7.3), set out to verify the 'Theory of Sea-Floor Spreading' as proposed by H.H. Hess (1962) and Robert S. Dietz (1962). It brought about an earth science revolution. The tremendous achievements of the Project in the 1970s owe much to the contributions from sedimentology. The following are a few representative examples of such sedimentological achievements.

**Fig.** 7.3 D/V *Glomar Challenger* (*reproduced by permission of DSDP*).

### 7.3.1 Walther's Law in the ocean-floor sediments

DSDP Leg 20 survey in 1971 confirmed that Walther's Law, originally established on terrestrial and shallow-water sediments (*see* Chapter 5.2), is applicable to sedimentation on the ocean floor (Heezen *et al.*, 1973; Hesse *et al.*, 1974; Okada and Kobayashi, 1974).

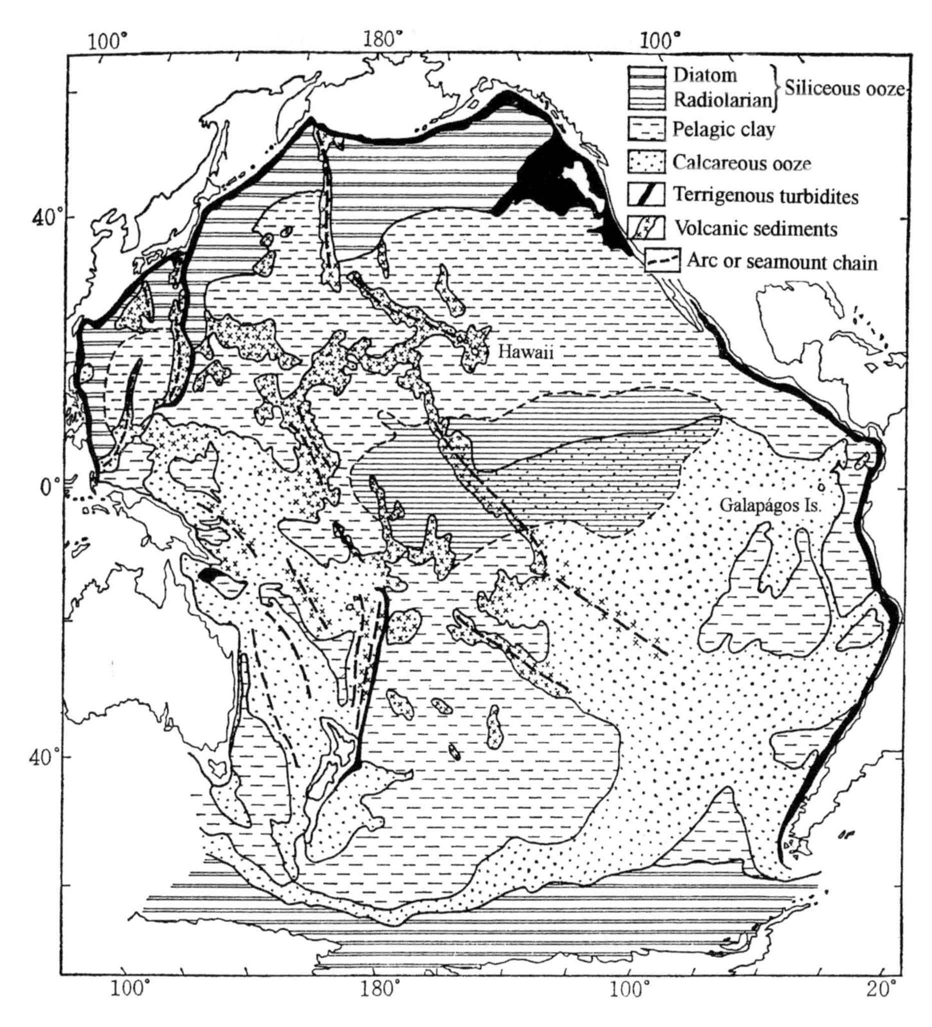


**Fig.** 7.4 Distribution of oceanic sediments in the Pacific Ocean (Okada and Kobayashi, 1974).

The distribution of modern deep-sea sediments in the Pacific Ocean reveals the characteristics shown in Figure 7.4. For example, calcareous ooze is distributed in the regions parallel to the equator and in areas shallower than the depth of 3600 m such as the East Pacific Rise. This is because the water depth of 4000 m is the average CCD thereabouts and on the floors shallower than this depth calcareous oozes are deposited. In addition, the reason why calcareous sediments can be deposited in the equatorial areas deeper than the CCD is explained as follows. High biogenic productivity in the equatorial zone due to upwelling of nutritious deep waters overwhelms the dissolution rate of calcium carbonates, and

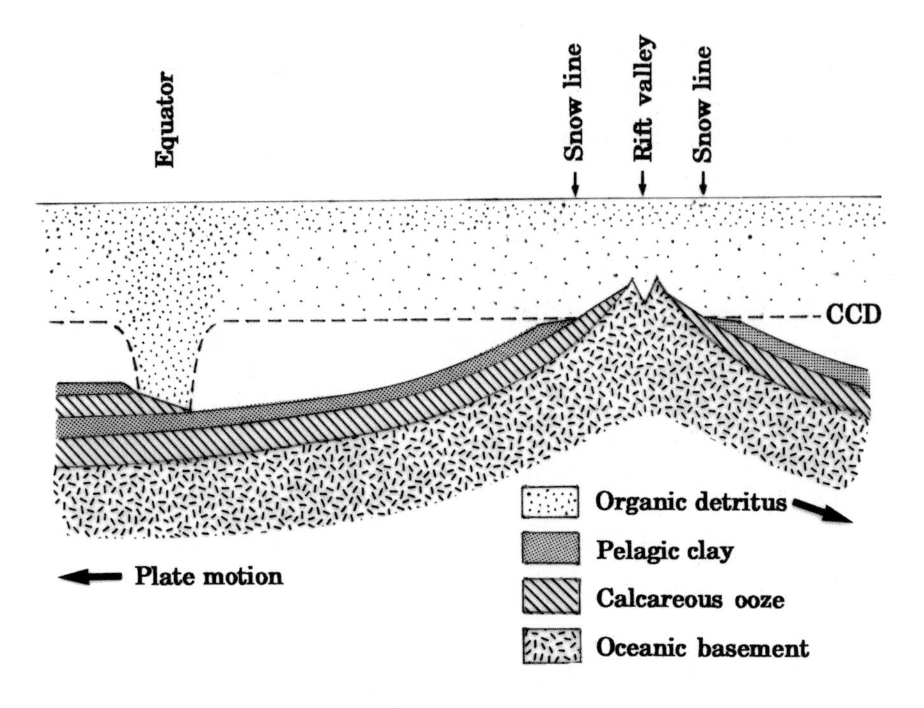


Fig. 7.5 Pelagic sedimentation on the moving oceanic plate (modified after Heezen *et al.*, 1973). (*Reproduced by permission of* Nature.)

considerable amounts of calcareous planktonic remains are thus deposited (*see* Fig. 7.5); that is, the CCD level is significantly deeper in areas of high biogenic productivity.

Seawaters containing siliceous components also show a silica compensation depth due to the solution of siliceous materials in silica-undersaturated sea waters. Considerable amounts of undissolved siliceous planktonic tests are deposited on ocean floors of 6000 m depth when the productivity is high. This is why siliceous oozes are deposited in high-latitude sea areas with high diatom productivity and in the equatorial zone also with high radiolarian productivity. In other deep-sea areas brownish pelagic clays are predominant. In addition, terrigenous turbidites characterize the trench zone, and volcaniclastic sediments are characteristic of the areas near island arcs and around seamounts.

The relationship between the areal distributions of these lithofacies and their vertical successions is listed here. According to the observations of cores drilled by DSDP Leg 20 and geophysical data, the following five lithologic units in descending order are recognizable in the deep-sea sediments of the Northwestern Pacific (Fig. 7.6):

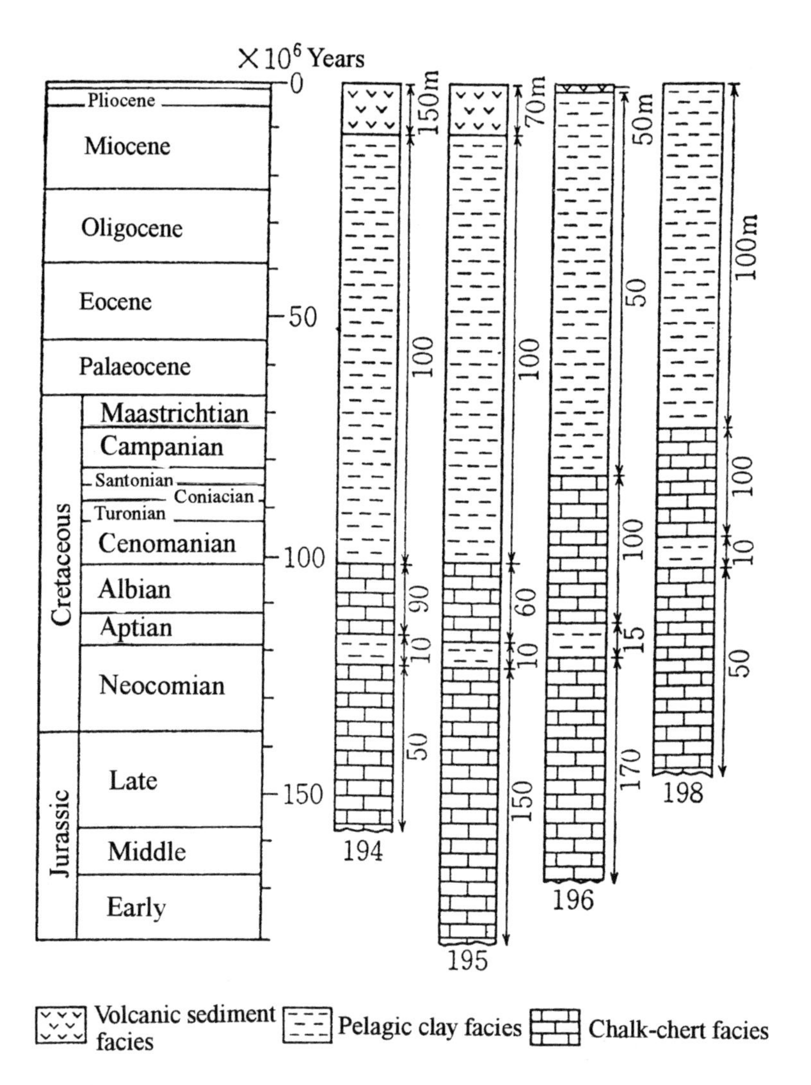


**Fig. 7.6** Lithostratigraphy of the deep-sea sediments in the West Pacific (Okada and Kobayashi, 1974). The numbers on the right side of each columnar section indicate the thickness in metres.

(1) Volcanogenic clayey silt (maximum about 150 m thick)

(2) Upper brown clay (50~100 m thick)

(3) Upper chert and chalk sequence (60~100 m thick)

(4) Lower pelagic clay (about 10~50 m thick)

(5) Lower chert and chalk sequence (about 150 m thick).

The uppermost sequence (1) is a non-calcareous, tuffaceous deposit composed mainly of diatom-rich clayey silt. In particular, the topmost few metres is rich in white pumice fragments. Intercalated thin layers of tuff contain volcanic glasses with a refractive index of about 1.512 and with $SiO_2$ contents of 70~80%, from which we can infer acidic volcanic activity.

The second sequence (2) is characterized by brown-coloured, homogeneous zeolitic clay rich in manganese micro-nodules and iron-oxides and sometimes bearing radiolarians.

The third sequence (3) is characterized by chalk beds associated with chert layers, which sometimes appear lenticular or nodular.

The fourth and fifth units are inferred from seismic data: drilled cores were not available.

This lithostratigraphy can be correlated to the seismic stratigraphy of the western North Pacific. That is to say, Horizon A' or the upper opaque layer and Horizon B' or the lower opaque layer defined by Ewing *et al.* (1968) are correlated to the upper and lower hard, dense chert and chalk layers, respectively.

It is important to point out that the boundary between adjacent lithofacies shows a diachronous relationship to chronological surfaces. Namely, the five lithofacies recognized in the Northwestern Pacific reveal that the upper and lower chert and chalk sequences get closer to each other towards the southeast, amalgamating into the pelagic clay facies and the chert and chalk facies sequence, and finally change into a single facies of calcareous ooze on the East Pacific Rise (Fig. 7.5).

Such a phenomenon can be explained in terms of the plate movement of the ocean floor. Namely, the siliceous and calcareous oozes or the lower chert and chalk facies sequence resting directly on the basalt basement were deposited on the East Pacific Rise immediately after its creation and as shallow as 2600~2700 m. Calcareous planktonic organisms could be deposited abundantly there, because the site was shallower than the CCD. The East Pacific Rise, therefore, is the place of deposition of the single lithofacies of siliceous and calcareous oozes (Fig. 7.5).

As the ocean floor gradually moves away from the rift axis of the ocean ridge, the ocean floor gets deeper. This is because the basaltic rocks of the submarine crust become heavier as they cool. Sclater *et al.* (1971) estimated the subsidence rate of the ocean floor as 65 m/m.y. This value suggests that, starting from an initial elevation of 2600~2700 m, which is the mean depth of the present East Pacific Rise, it takes about 20 million years for the ocean crust to subside below the depth of 4000 m. If the rate of sedimentation of oozes is estimated as

10 mm/1000y during this period, they attain the maximum thickness of 200 m. This thickness is in harmony with the actual measured thickness of the sediments. When the ocean crust moves down below 4000 m, as the water depth exceeds the CCD, biogenic oozes can no longer be deposited and pelagic clays become predominant. The sediments thus deposited are the lower pelagic clay. It covers siliceous and calcareous oozes deposited directly on the ocean-floor basement, and protects them from further solution. As a consequence, it follows that the ocean floor before reaching the equatorial zone is covered by two lithofacies, that is – the lower chert and chalk sequence and the lower pelagic clay sequence (Fig. 7.5).

When the moving ocean crust enters the equatorial zone, biogenic oozes are again deposited due to the apparent deepening of the CCD to the ocean floor as the nutritious upwelling deep waters cause high productivity of planktons. These sediments constitute the upper chert and chalk sequence, thus creating the three lithofacies of sediments in the equatorial zone (Fig. 7.5).

Once the equatorial zone is crossed, abyssal clay sedimentation predominates, because the CCD rises as the supply of planktonic tests decreases. This sediment overlies the chert and chalk sequence, which was deposited in the equatorial zone, showing the four lithofacies revealed by the drilling on DSDP Leg 20 (Fig. 7.5). Afterwards, volcanogenic sediments derived from island-arc volcanism related to plate subduction at the northwest edge of the Pacific basin are characteristic of the uppermost sequence of the deep-sea stratigraphy (Fig. 7.6, 7).

This stratigraphy is completed by the sedimentation of terrigenous turbidites at the top when the oceanic plate approaches the trench zone.

The stratigraphic relationship constructed by such a dynamic relationship between plate movements and sedimentation is called plate stratigraphy (Berger and Winterer, 1974). In this context, a coarsening-upward sedimentation characterized by a unit sequence of lower oceanic sediments and upper turbidites was clearly exhibited for the first time in the drilled cores from the Nankai Trough, off Shikoku, Southwest Japan (Moore and Karig, 1976). This became the fundamental stratigraphy in the island-arc accretionary zone for understanding the complicated geological structures in ancient active margins (Lash, 1985). Sedimentology has thus contributed a great deal to the development of plate stratigraphy in the modern ocean floor and to the demonstration of sea floor spreading.

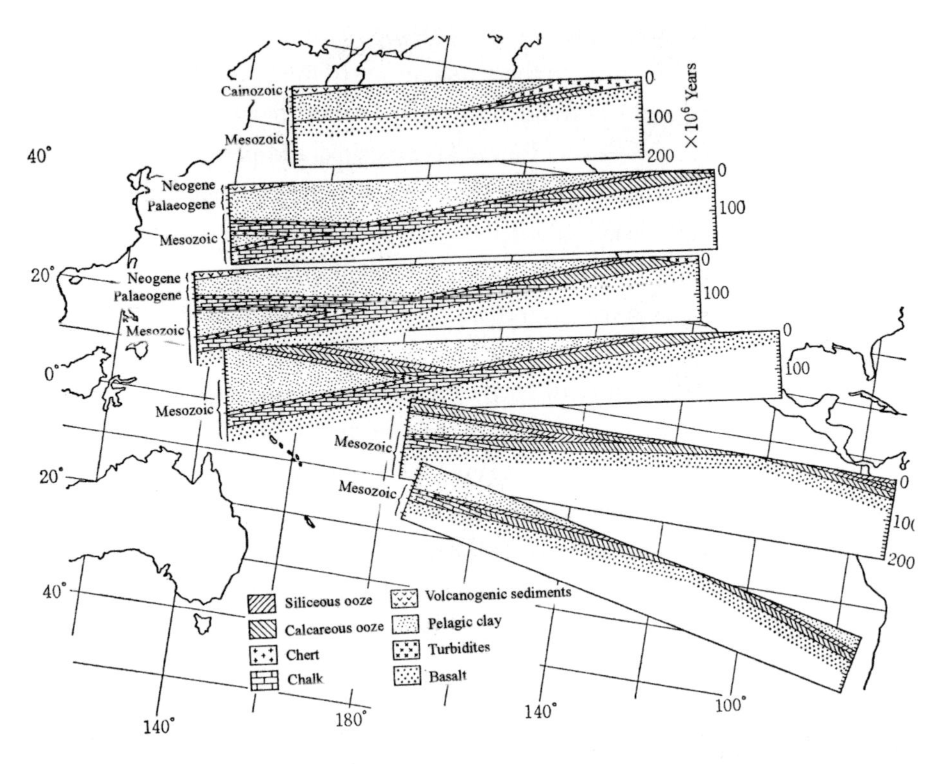


Fig. 7.7. Vertical and horizontal changes of lithofacies of the Pacific oceanic sediments in relation to oceanic plate movement (modified after Heezen *et al.*, 1973). The numbers on the right of each section give time in millions of years.

## 7.3.2 The desiccated Mediterranean

Deep-sea drilling in the Mediterranean Sea was twice carried out by D.V. *Glomar Challenger*: DSDP Leg 13 in 1970 and Leg 42A in 1975. On both cruises, Kenneth J. Hsü (Fig. 7.8), sedimentologist and professor of the Swiss Federal Institute of Technology in Zurich, played an important role as one of the co-chief scientists.

At that time, a strong acoustic reflector of unknown origin called the M-reflector (so called after the initial letter of 'Mediterranean') beneath the floor of the Mediterranean had been well known to geophysicists. Through the cores from the Mediterranean, however, the shipboard scientists discovered that the M-reflector consisted of widespread evaporites. In particular, the discovery of the lithologies and the stratigraphy of the evaporites at six sites in the Balearic Basin and the Tyrrhenian Basin in the western Mediterranean was important. Fig. 7.9 shows the

Fig. 7.8 Kenneth J. Hsü (photo by H. Okada in 1993).

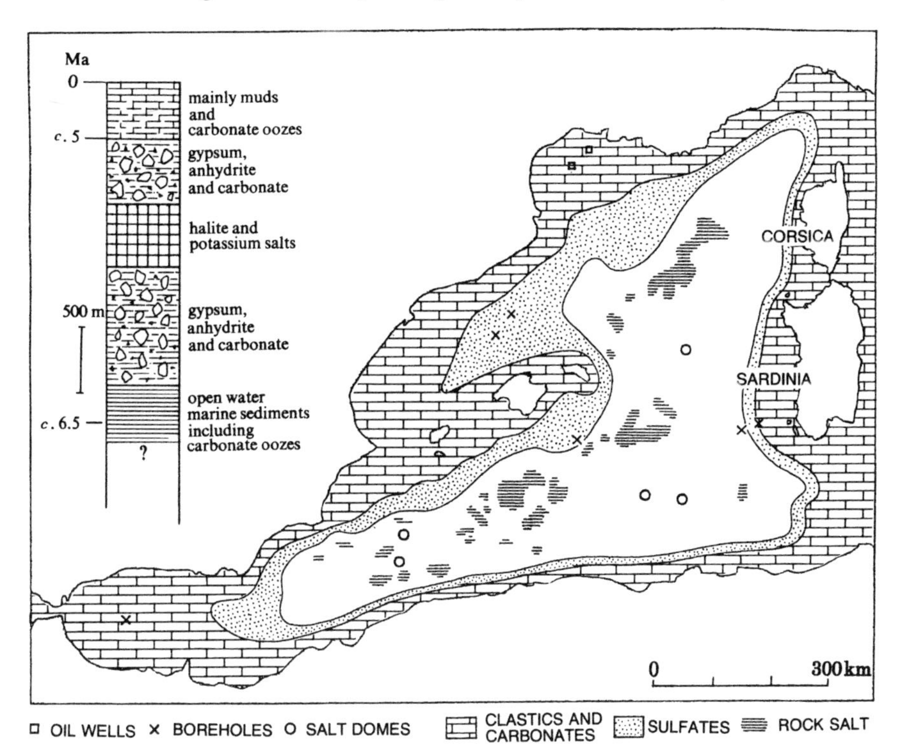


**Fig. 7.9** Distribution of evaporite facies in the Balearic Basin in the western Mediterranean Sea (after Hsü, 1972b). Generalized succession of the late Miocene and younger Mediterranean submarine sediments on the left (Open University, 1978). (*Reproduced by permission of Donald G. Garber.*)

distribution of evaporites in the Tyrrhenian Basin (Hsü, 1972a, b) and the general stratigraphy of the bottom sediments in the Mediterranean in the upper left of the figure (Open University, 1978). It was the stromatolitic evaporites to which Ken Hsü paid a special attention. He considered that anhydrites were deposited from saline water in a hot and arid basin, like a sabkha, at temperatures above 35°. Consequently, he developed his idea that an originally deep Mediterranean Sea dried out in the late Miocene and thus the evaporite formation was deposited. This he explained as follows (Hsü, 1972a, b, 1983; Hsü *et al.*, 1973).

From 150 Ma to 20 Ma ago the Mediterranean Sea was a part of the Tethys Ocean, which acted as a channel through which warm and saline tropical currents flowed and dense, deep waters were transported westwards into the Atlantic Ocean. In the early Miocene the Arabian Plate collided with the European Plate, and the Mediterranean separated off from Tethys and the proto-Pacific Oceans to the east. It was connected with the Atlantic Ocean only through a narrow strait in the west. The separation of the eastern marine connection with the Indian Ocean gradually developed drier climatic conditions in the Mediterranean region. Finally, during the short time from 6.5 Ma to 5.5 Ma, at which time the Mediterranean was separated from the Atlantic Ocean by a barrier that prevented replenishment of evaporated water, consequently large amounts of evaporites were precipitated in the Mediterranean Basin, and very quickly it dried up completely. The sedimentation process of the Mediterranean evaporites is indeed well explained by the evaporating-dish model or the 'bull's eye' model of evaporite deposition (Fig. 7.10). This short time event, which caused the evaporite precipitation and complete drying up of the Mediterranean, is called the Messinian salinity *crisis*, because it had brought about major environmental and climatic changes (Hsü *et al.*, 1977). Consequently, the circum-Mediterranean land areas were subjected to an extraordinary fall of water tables that reinforced the activity of groundwater, and this is considered to have promoted the formation of extensive caverns in limestone areas – called the *karst* in Yugoslavia.

There is a different view for the origin of the Messinian evaporites: that they were deposited from stagnant hypersaline waters that had restricted circulation due to sea level fall but retained an exchange with the Atlantic Ocean (Sonnenfeld, 1985; Sonnenfeld and Finetti, 1985). Incidentally, as a cause of significant global sea-level fall in Messinian times, Minoura and Hisajima (1993), based on their studies of detailed lithostratigraphy, biostratigraphy and chemostratigraphy of Miocene deposits in the Sorbas Basin of the Iberian Peninsula, pointed to the importance of the large scale development of contemporaneous Polar ice sheets.

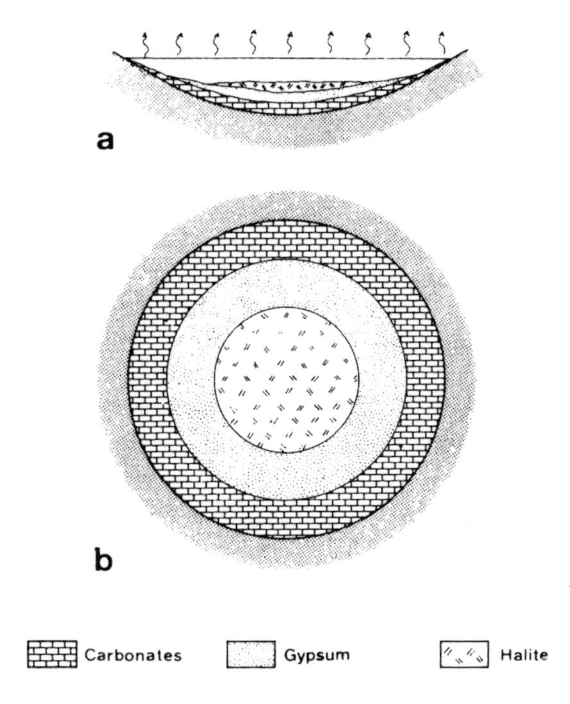


**Fig. 7.10** An evaporating-dish model or "bull's-eye" pattern of evaporite deposition (Hsü, 1972a). The first salt to precipitate is carbonate (limestone or dolomite). As the water level drops and the brines become concentrated to 1.10 g/cm$^3$, gypsum precipitates; at 1.13 g/cm$^3$, anhydrite would be deposited, and at 1.215 g/cm$^3$ finally in the centre of the deepest place of a salt pan, halite would be precipitated. (*With permission of Elsevier.*)

The idea and the establishment of the theory that 'The Mediterranean Sea was a desert' came from Kenneth J. Hsü, the co-chief scientist of the DSDP Leg 13 on board *Glomar Challenger* to the Mediterranean Sea and are vividly described by Hsü (1983, 1992). In these books Hsü enthusiastically pointed to the importance of sedimentology (Takayanagi, 1999).

### 7.3.3 Bottom currents and their sediments

Only after the 1950s was it realized that the waters of the deep ocean floor are not always still, but are subject to fairly strong currents with velocities of more than 10 cm/sec. Near-bottom ocean currents naturally play a significant role in the processes of erosion of submarine sediments and their resedimentation. In this section the geological significance of bottom currents and their relationship to turbidity currents are discussed.

### *Bottom currents and their circulation system*

It was believed for a long time that the deep-sea floor is a placid environment where sediments are constantly and continuously deposited. However, as the results of the Deep Sea Drilling Project accumulated, the existence of regional erosion at the deep ocean floor was recognized (Johnson, 1972; Kennett *et al.*, 1972; Edgar *et al.*, 1973; Edwards, 1973; Rona, 1973; Davies *et al.*, 1975; Langseth, Okada *et al.*, 1978; Vallier, Thied *et al.*, 1979; Hussong, Uyeda *et al.*, 1982), and the importance and origin of such abyssal bottom currents have been discussed. The geological significance of bottom currents is considered here.

Surface ocean waters tend to make vertical circulations of water masses according to changes of temperature and salinity, and these phenomena are called thermohaline circulations. Such ocean circulations are generally shown by the movement of water masses along the ocean floor which are generated because the surface water masses in higher latitudes are cooled and descend. These waters flow as currents along the bottom topography due to the density stratification of deep waters. In other words, they flow parallel to depth contours. Therefore these currents are often called contour currents. They are also called geostrophic contour currents, because their movement is governed by the Coriolis Force, the product of the centrifugal effect caused by the rotation of the Earth.

The presence of bottom currents able to transport sediments in the deep ocean was first suggested by George Wüst, a German geophysical oceanographer, as early as 1936. He estimated the maximum velocity of bottom currents in the Atlantic to be 10~15 cm/sec. Influenced by his work, C. D. Hollister submitted his Ph.D. dissertation on the geological effects of deep circulation in the Western North Atlantic to Columbia University in 1967, and this became the basis of the new concept of 'contourites' proposed by Heezen *et al.* in 1966 (Hollister, 1993).
Clear evidence of the existence of bottom currents was furnished first by H. W. Menard in 1952 when he photographed ripple marks at a depth of 1500 m on the flanks of the Sylvania Seamount in the South Pacific. Since then, much data on ripple marks, erosive structures and linear structures produced by current flows on the deep ocean floor have been reported. In addition, based on the study of grain-size distribution of modern coarse-grained clastics on the bottom of the North Atlantic, J. F. Hubert (1964) criticised the general trend of concluding that deep-sea sands had a turbidity current origin and pointed to the importance of thermohaline bottom currents rather than turbidity currents as a transporting mechanism.

As to the origin and circulation of ocean bottom currents, H. Stommel (1958) was a pioneer in recognizing a global current circulation system of bottom

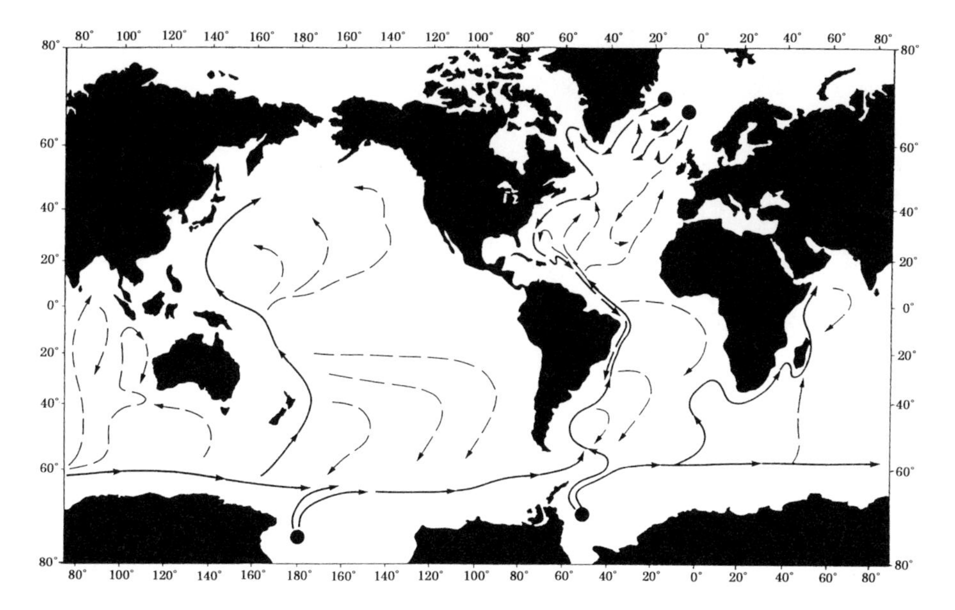


**Fig. 7.11** The deep circulation pattern of bottom currents (Stow and Lovell, 1979). (*Reproduced by permission of Elsevier.*)

currents deeper than 2000 m. D. A. Stow and J. P. B. Lovell (1979) later modified Stommel's circulation patterns (Fig. 7.11). Recently, J. C. Faugères *et al.* (1993) demonstrated deep circulation patterns in the Atlantic, as shown in Figure 7.12. These major circulation systems are separated into three main bottom currents: the North Atlantic Deep Water (NADW), the Western Boundary Undercurrent (WBUC) and the Antarctic Bottom Water (AABW), all of which are part of a global circulation. The NADW is generated in the Greenland Sea and the Norway Sea and flows southwards along the western margin of the North Atlantic. The AABW originates in the Weddell Sea of the Antarctic Ocean and flows northwards in the western part of the South Atlantic. In the sea areas where the NADW and the AABW encounter each other, the former flows southwards on the land side, whereas the latter flows northwards on the ocean side (Fig. 7.13). Although bottom currents generally flow along the ocean floor at low velocities (less than 2 cm/sec), they always accelerate along the western margins of every ocean basin due to the effect of Earth's rotation (Stommel and Aarons, 1960; Rona, 1973). This is the reason why they are called the Western Boundary Undercurrent (WBUC).

It is noteworthy that W. V. Sliter (1977) has been able to present the palaeocirculation patterns of bottom currents in the southern hemisphere in the Late Cretaceous (Fig. 7.14).

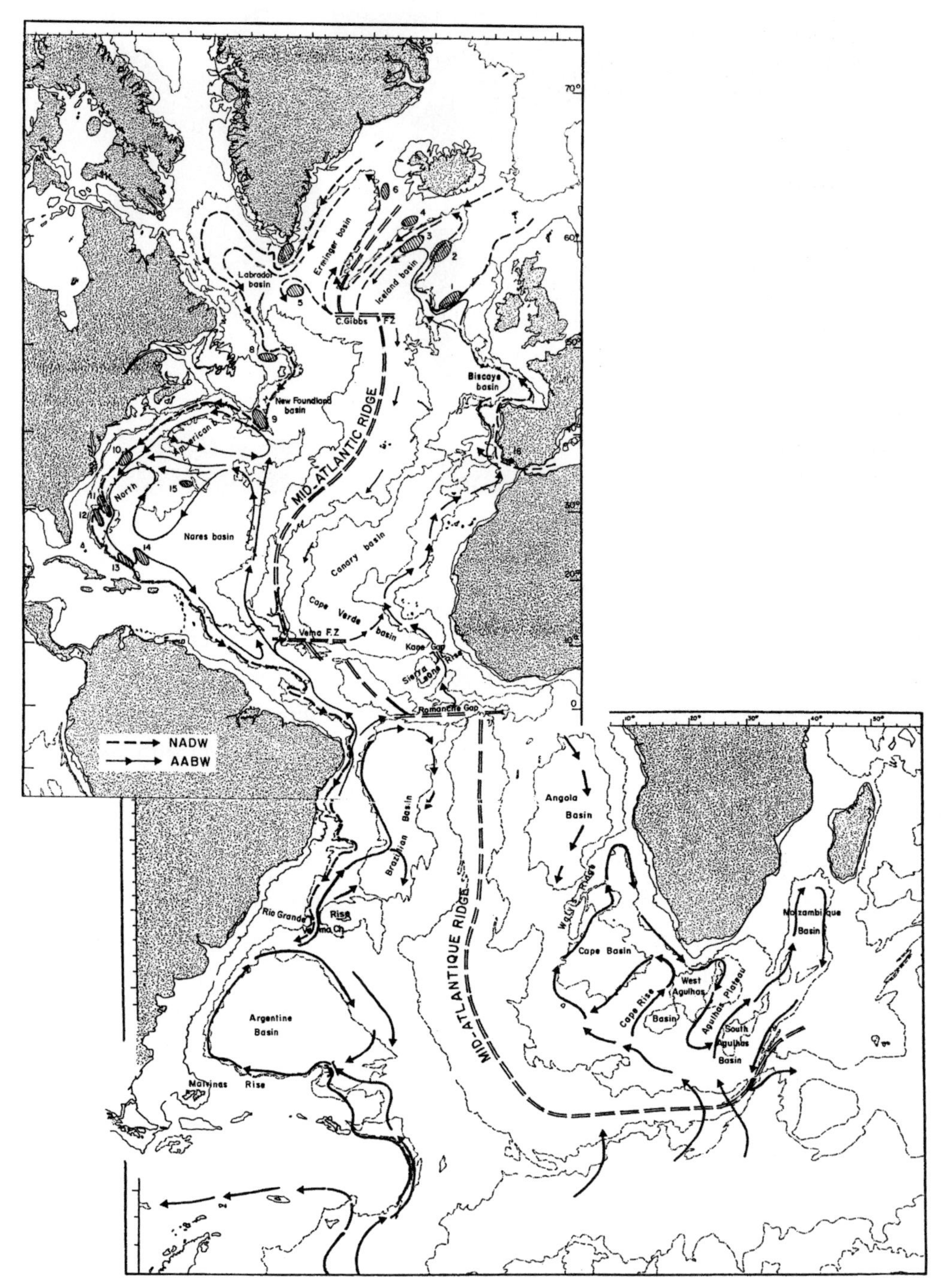


**Fig. 7.12** The deep circulation pattern in the Atlantic Ocean (Faugères *et al.*, 1993). NADW: North Atlantic Deep Water, AABW: Antarctic Bottom Water.

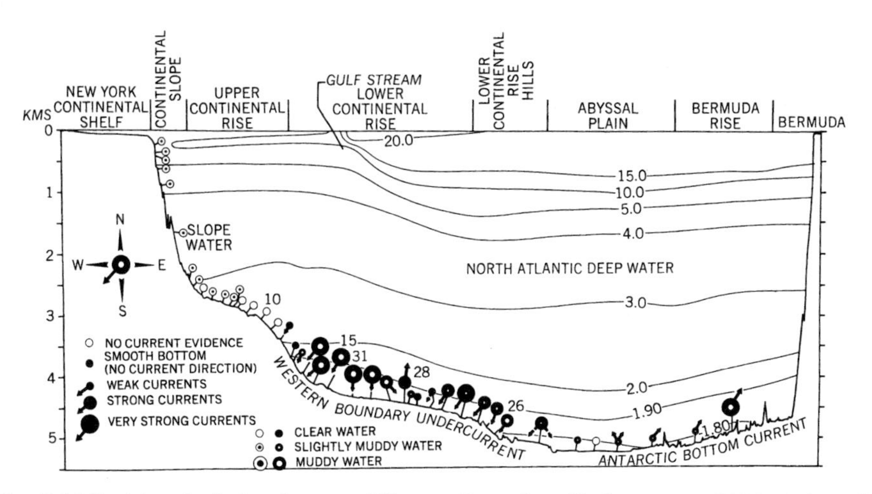


**Fig. 7.13** Positional relation between Western Boundary Undercurrent: WBUC (North Atlantic Deep Water: NADW) and Antarctic Bottom Water: AABW (Heezen and Hollister, 1971).

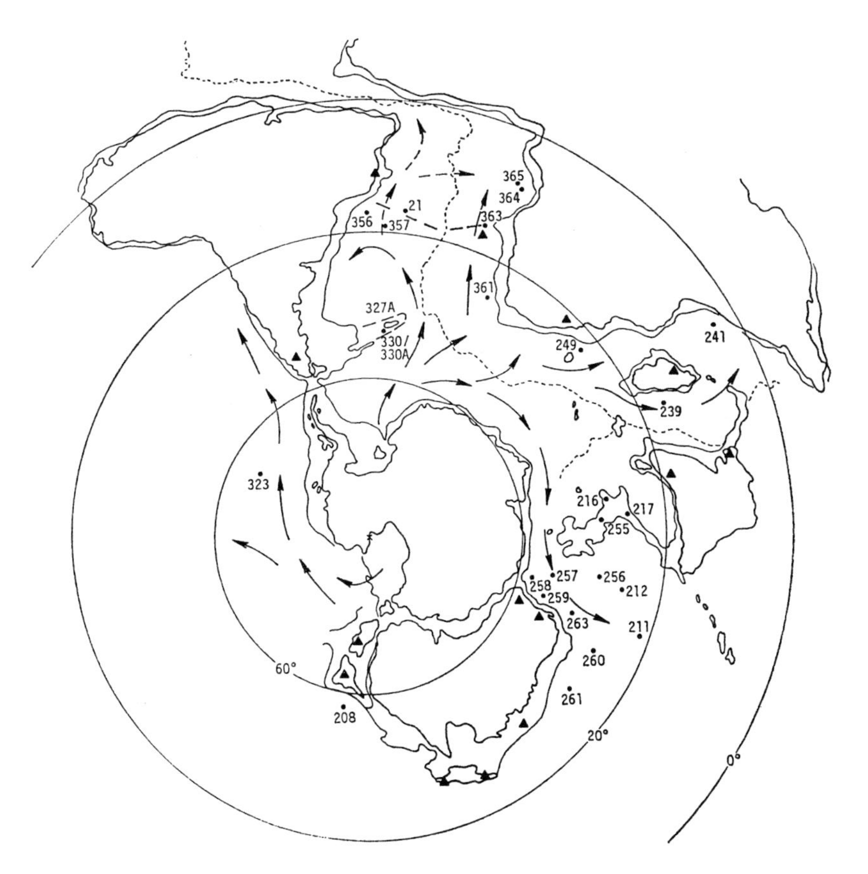

**Fig. 7.14** A circulation pattern of bottom currents in the Late Cretaceous (Sliter, 1977). Figures indicate the DSDP drilling sites. (*Reproduced by permission of DSDP.*)

*Velocity of bottom currents*

As stated above, bottom currents generally show quite low velocities, but they obtain higher velocities in the Western Boundary Undercurrent and in constricted topographic depressions. Observed data indicate that the velocity of the WBUC attains, in general, 10 to 30 cm/sec (Table VII-1). Luyten (1977) carried out a continuous measurement of bottom current velocity for eight months on the upper continental rise off Cape Cod, Massachusetts, and recorded a westward flow, with 5 cm/sec on average, along the topographic contour with pulsated flows at 40 cm/sec. Betzer *et al.* (1974) also measured similar pulsating bottom currents with highest short-lived velocities from 15 to 47 cm/sec beneath the Gulf Stream near Cape Hatteras. Clearly, it may be concluded that the velocities of bottom currents can vary from a few cm/sec to more than 40 cm/sec, and sometimes over 100 cm/sec (Table 7.1).

**Table 7.1** Measured velocities of bottom currents (edited from Heezen and Hollister, 1964; Betzer *et al.*, 1974; Stow and Lovell, 1979).

| Sea area | Depth (m) | Velocity (cm/sec) |
|---|---|---|
| ***WBUC*** | | |
| SE of Iceland | 2100 | 20 ~ 30 |
| Greenland-Iceland-Faeroes | 2000 ~ 3000 | 0 ~ 12 |
| Iceland-Faeroes Ridge | ? | 20 ~ 30 (109) |
| Off Blake Plateau | 3300 ~ 3500 | 9 ~ 18 (20) |
| Antillean-Caribbean Basins | 4000 ~ 8000 | 10 |
| West Bermuda Rise | 5200 | 12~17 |
| Blake Bahama Outer Ridge | 4300 ~ 5200 | 6 (26) |
| Rise off New England | 3000 ~ 5000 | 10 ~ 20 (26.5) |
| Greater Antilles Outer Ridge | 5300 ~ 5800 | 2~10 (20) |
| SE of Cape Hatteras | 1265 | 0.3 (15) |
| | 2575 | 10.9 (47) |
| | 2810 | 12.6 (44) |
| | 3220 | 6.8 (23) |
| | 3720 | 0.8 (26) |
| | 4145 | 9.1 (16) |
| W of Bermuda | 4400 | 42 |
| S of Bermuda | 4400 | 7.6 |
| ***AABW*** | | |
| Tonga Trench | >4800 | 5 ~ 15 (19) |
| E of Madagscar | 4500 | 6 ~ 7 |

Velocities in parentheses indicate maximum velocities measured.

### *Erosive power of bottom currents*

Generally speaking, cohesionless sand grains are more easily moved than clay and silt particles, and begin to move at a current velocity of 20 cm/sec (Hjulström, 1939; Sundborg, 1956). It is not surprising, therefore, that bottom currents generate ripple marks on sandy, deep ocean floors. Even adhesive oozes and clays show erosional structures – when current velocities are 7–10 cm/sec on calcareous oozes (Southard *et al.*, 1971); 30–100 cm/sec on red clays (Lonsdale and Southard, 1974), and 7.5–22.3 cm/sec on fine-grained sediments (Young and Southard, 1978). Coarse-grained *Globigerina* oozes need a current velocity of 30 cm/sec to be eroded (Heezen and Hollister, 1971, p. 358).

Evidence of ocean-bottom deposits being eroded by bottom currents over wide areas and for long periods is well known in the West Pacific. For example, Hollister *et al.* (1974) and Hussong, Uyeda *et al.* (1982) demonstrated large-scale regional erosion in the Northwest Pacific between the Shatsky Rise and the Caroline Islands, which created a major sedimentary hiatus in the period from the Cretaceous to the Quaternary (Fig. 7.15). They ascribed the origin of the

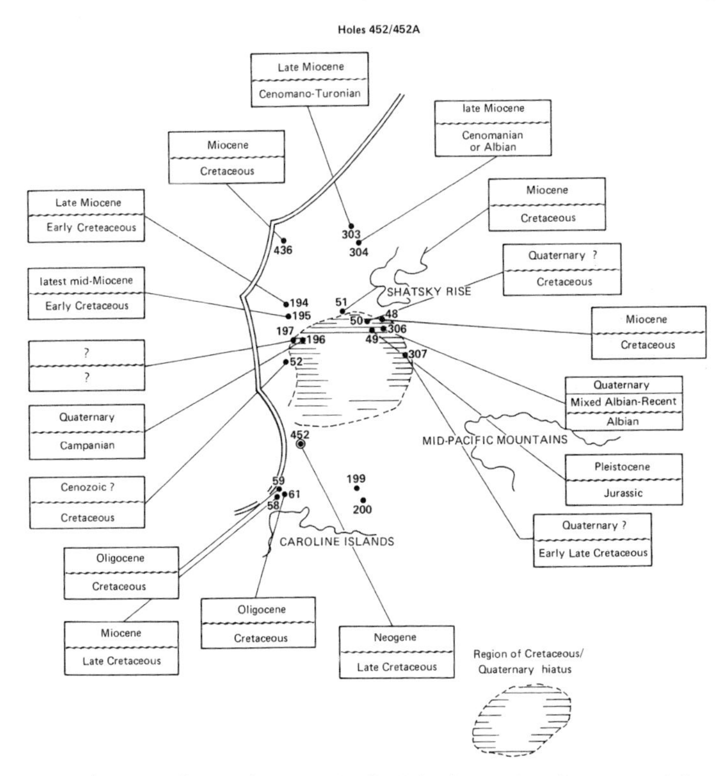


**Fig. 7.15** Distribution of great hiatuses in the Northwest Pacific Ocean (after Hussong, Uyeda *et al.*, 1982). Figures indicate the drilling sites of DSDP. (*Permission of DSDP.*)

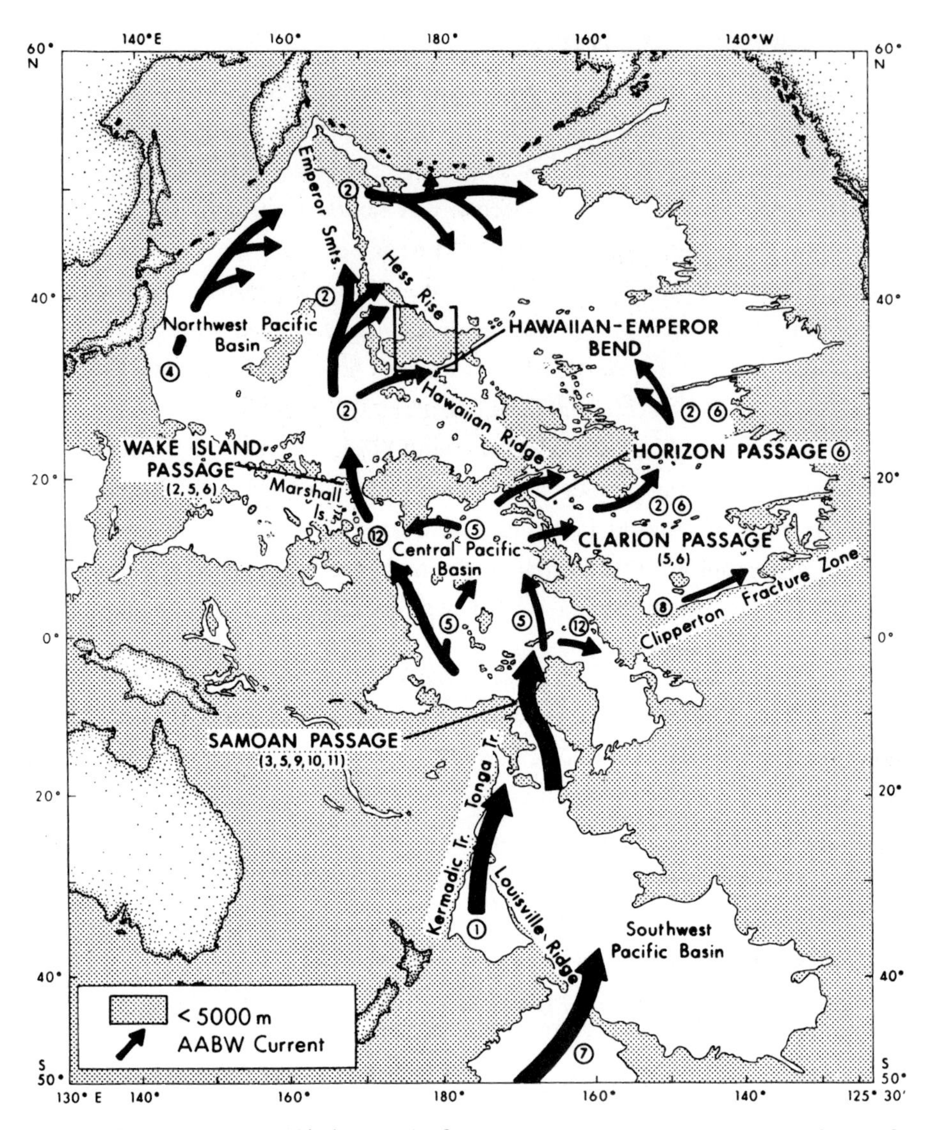


**Fig. 7.16** Current patterns (dark arrows) of Antarctic Bottom Water (AABW) in the Pacific (after Nemoto and Kroenke, 1981). (*By permission of the American Geophysical Union.*)

erosion to the AABW. In fact, this ocean area is strongly controlled by a branch of the AABW, as Nemoto and Kroenke (1981) reported (Fig. 7.16).

### *Contourites*

Bottom currents are, in general, observed in areas of deep-water and three-dimensionally well defined topography. These places are well represented by the lower part of the continental shelf and the continental rise, where bottom

**Fig. 7.17.** Bruce C. Heezen on board *Glomar Challenger* (DSDP Leg 20) in September, 1971 (photo by H. Okada). Bruce Heezen died aboard a research submarine south of Iceland in 1977.

currents show a velocity of more than 40 cm/sec, sometimes even more than 100 cm/sec (Table 7.1). Thus it has become clear that bottom currents have enough power to erode sands deposited by turbidity currents and to resediment them. It has been noted that many beds of well-sorted, clean sands have well developed ripple marks. Bruce C. Heezen (1924–1977) (Fig. 7.17), former professor of the Lamont-Doherty Geological Observatory, Columbia University, who studied such sedimentary structures, defined the sands and silts deposited by contour currents as contour-current deposits or contourites (Heezen *et al.*, 1966).

According to Heezen *et al.* (1966) and Bouma (1972a, b), contourites show the following features:

(1) Always well sorted: the muddy matrix component is less than 5%.

(2) No graded bedding: both the bottom and the upper surfaces of the bed are sharp.

(3) The upper surface of contourite beds is characterized by ripple marks, and the internal structures by cross lamination.

(4) Thin beds: generally less than 5 cm thick.

Figure 7.18 shows an exposure of thin contourite layers intercalated with thick beds of turbidite sandstones. Stow and Lovell (1979) and Stow (1985) also presented features of contourites. Daniel J. Stanley (1993) showed a continuous lithofacies model from turbidite beds to contourite beds (Fig. 7.19).

**Fig. 7.18** An example of ancient contourites from the Eocene Kayo Formation at Nago, Okinawa Island, Japan (photo by H. Okada). Note two contourite layers (5 to 6 cm thick, cross-stratified) intercalated between thick turbidite sandstones.

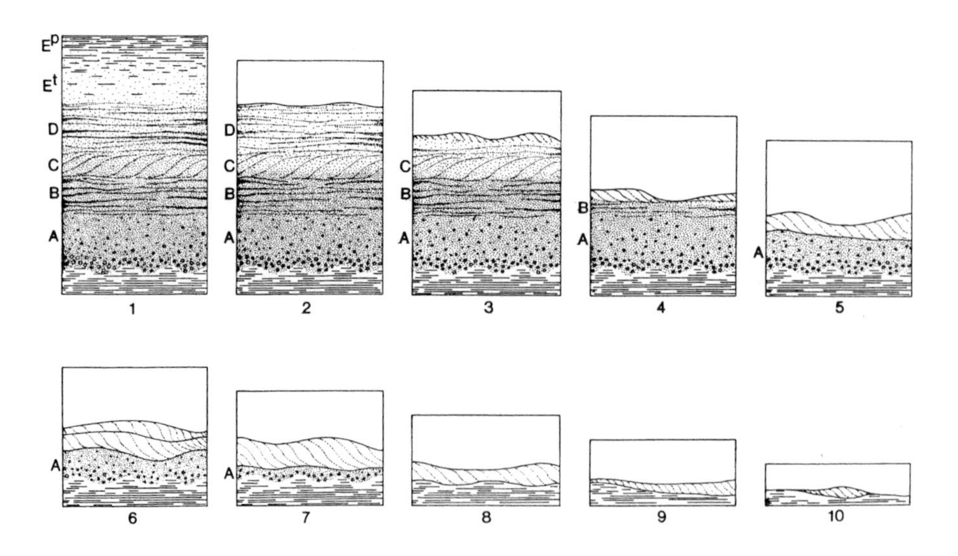


**Fig. 7.19** A turbidite-to-contourite continuum model (Stanley, 1993). A–E: Bouma sequence of a turbidite. Contourite stratal types 8 to 10 are the reworked end-products of a turbidite. (*Reproduced by permission of John Wiley & Sons Inc.*)

### *Bottom currents of highly variable velocities and flow directions observed in Suruga Bay and Sagami Bay, Japan*

As stated above, general features of bottom currents may be understood as simple flows along the contour lines of deep-sea topography. However, it has been demonstrated that the bottom currents in deep bays tend to fluctuate rapidly in velocity and direction probably due to strong internal tidal waves in a very steep sided embayment such as in Suruga Bay and Sagami Bay, central Japan (Ohta, 1983; Okada and Ohta, 1993) (Fig. 7.20). In this repect, Takano and Hara (1970) reported an interesting observation on the behaviour of bottom currents at the depth of 700~1350 m in Sagami Bay. According to their observations (Fig. 7.21), bottom current velocities and directions of the water movements are highly variable and current speeds of more than 20 cm/sec are quite common. Ohta (1983) showed the changes in direction and velocity of bottom currents in Suruga Bay by observing the behaviour of an effective biological current indicator, the small deimatid holothurian *Peniagone japonica* Ohshima. Over a period of 60 minutes directions went through 360°: a complete turn with many changes and reverses (Fig. 7.22), and current velocities reached 20 cm/sec to 50 cm/sec.

Mitsuzawa *et al.* (1993) also carried out observations of both bottom currents and turbidity currents for 57 days from June to August in 1990 at deployment points at depths of 850 m, 1150 m, and 1960 m in the axial zone of the Suruga Trough where the water depth was 1975 m. As a result, it was demonstrated that the bottom current velocity and its azimuth of polarization are much stronger at the 1960 m depth than at the two upper depths, and at every depth the current velocity showed an approximately thirteen day cycle due to tidal forces (Fig. 7.23). In addition, at the spring tide, strong currents from the SSW direction with a period of a day and with the maximum velocity of about 60 cm/sec were measured. At the neap tide, the current velocity at the bottom point was below 20 cm/sec with a period of half a day. High velocity currents of over 40 cm/sec with polarization of a southern direction were also observed. These were mainly caused by internal tides related to the topographical conditions of the Suruga Trough with its north–south axis. Incidentally, a turbidity current caused by a typhoon in August 1990 showed a maximum velocity of over 70 cm/sec.

Additional data of the velocities of bottom currents was provided by Ohta (1989) who measured strong bottom currents of the order of 40 cm/sec at a height of several metres above the sea floor along the axis of the Suruga Trough. Okada (1988) observed northeastward bottom currents with a velocity of 50 cm/sec at a depth of 4200 m on the trench axis of the northernmost part of the

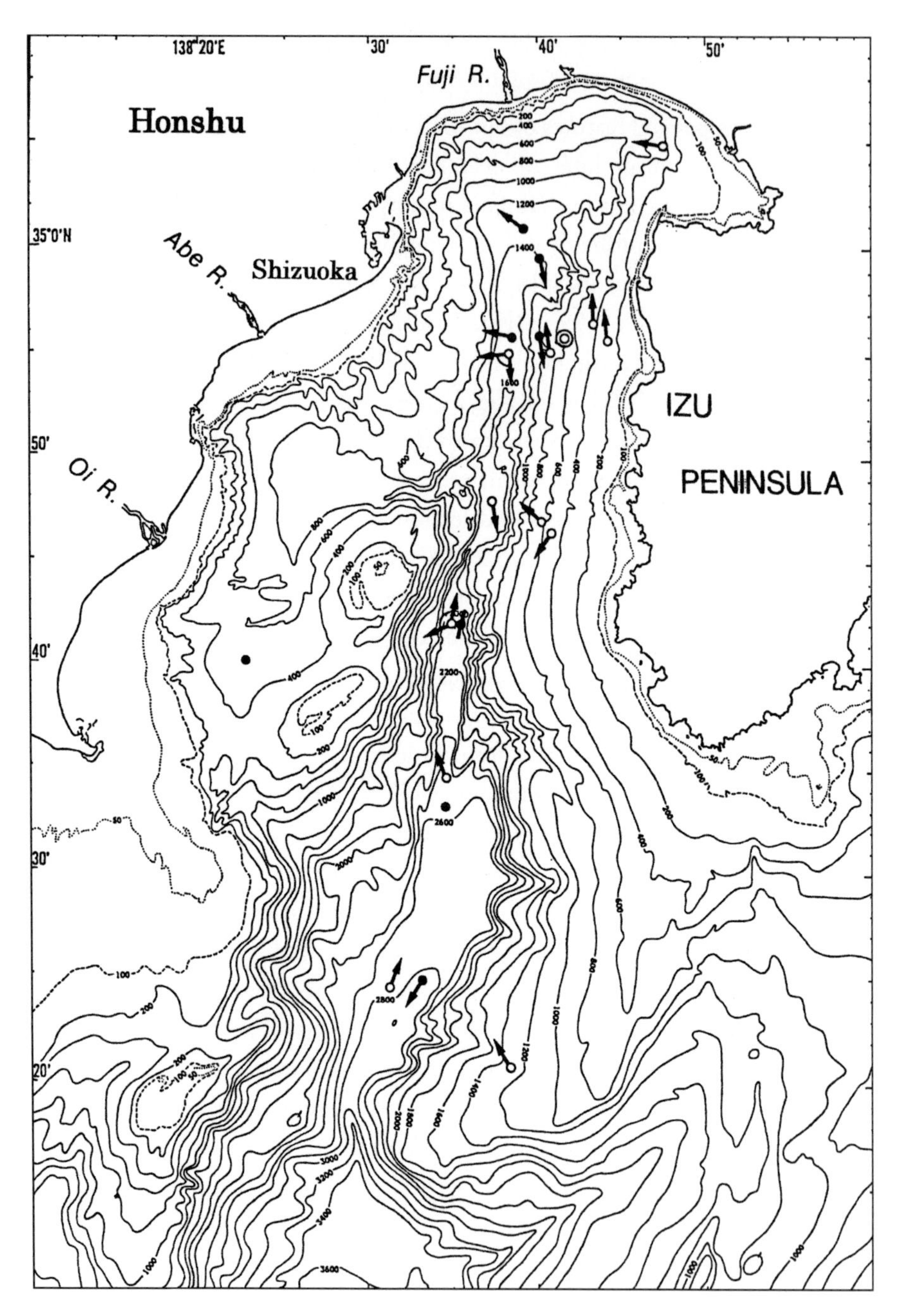


**Fig. 7.20** Bottom-current directions determined from the current-induced bottom markings (dots) and from the orientation of benthic organisms (circles) in Suruga Bay, central Japan (Okada and Ohta, 1993). Double arrows linked with an arc indicate that the directions of bottom current changed during the period of observation, and likewise, double circles represent a complete turn of the current direction within 40 minutes.

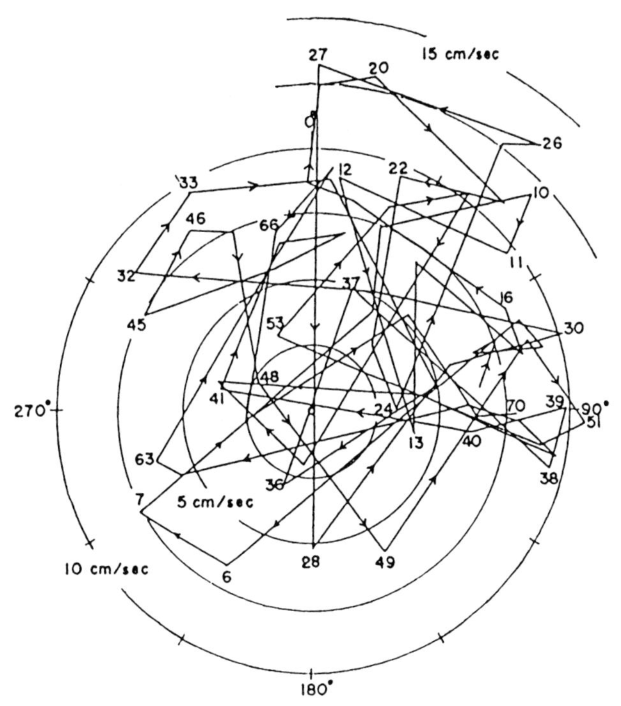


**Fig. 7.21** Hodograph of the velocity sector at the innermost part of the Sagami Trough at the depth of 1300 m showing highly variable velocities and directions of bottom currents (Takano and Hara, 1970).

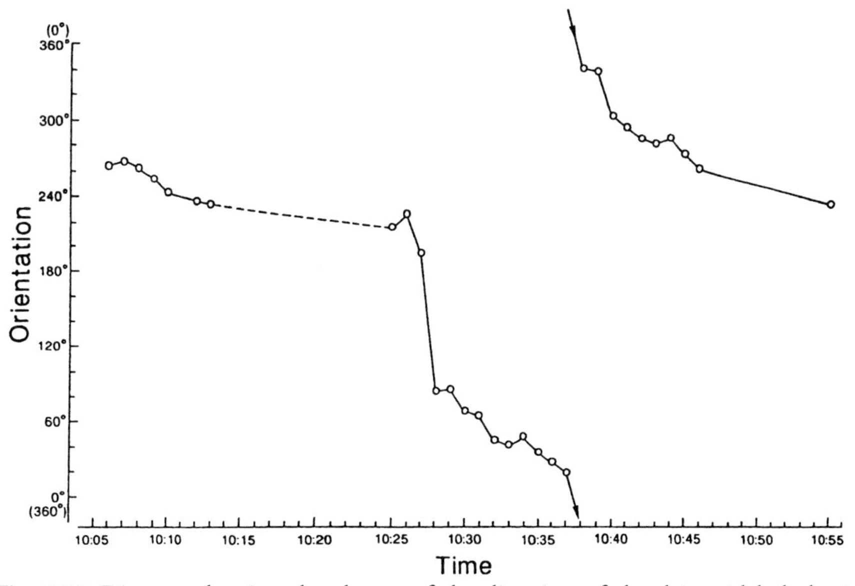


**Fig. 7.22** Diagram showing the change of the direction of the deimatid holothurian *Peniagone japonica Ohshima* at a depth of 646–698 m in Suruga Bay off Heda, Izu Peninsula, central Japan (*see* also Fig. 7.20) (Okada and Ohta, 1993). Note an abrupt change in current directions during only two minutes. The direction is represented as a mean direction.

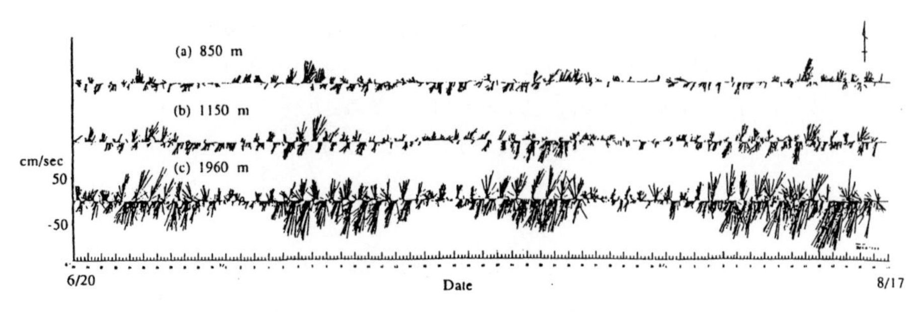


**Fig. 7.23** Observed current velocity time series at the 1975 m site in the Suruga Trough, central Japan, from 20 June to 17 August, 1990 (Mitsuzawa *et al.*, 1993). Note the approximately 13 day cycle in current velocities.

Nankai Trough, which is opposite to the southwestward slope direction of the trench axis. Taira and Teramoto (1985) measured extremely strong flows at the axial part of the Sagami Trough, a return flow of 36.9 cm/sec towards 298° after the peak of 49.0 cm/sec to 145°. The return flow lasted for three hours with a velocity of more than 35 cm/sec.

In connection with this subject, Gao and Eriksson (1991) and Gao *et al.* (1998) emphasized the importance of internal tides and internal waves as a cause of deep-water traction current deposits.

### *Bottom currents and nepheloid layers*

The existence of the near-bottom water in the ocean containing high concentrations of suspended sediments was first detected by M. Ewing and E. M. Thorndike (1965) through light-scattering studies of the water column from the surface to the bottom of the ocean. They called this near-bottom water the *nepheloid layer*. The thickness of nepheloid layers ranges from 55~820 m in the North Pacific (Ewing and Connary, 1970) to 300~2400 m in the North Atlantic (Eittreim *et al.*, 1969), and are estimated to be about 1000 m on average in all oceans. The concentration of materials suspended in nepheloid layers ranges from 0.44~0.56 mg/l compared to 0.03~0.24 mg/l for clear water (Jacobs, 1978). Suspended materials consist of terrigenous debris, planktonic bioclasts, pellets, and organic materials. Particles less than 2 $\mu$m size prevail ranging from 85 to 96% in nepheloid layers compared to 76~87% in the clear water.

The areas of nepheloid layers with high suspension and large thickness coincide with the areas where such bottom currents as the WBUC and AABW are strong (Ewing and Connary, 1970; Eittreim *et al.*, 1969). In the North Atlantic the densest part of nepheloid layers occurs near the boundary between the continental rise and the abyssal plain, and coincides with the axis of WBUC. This suggests that fine particles are suspended in waters for a long time by the bottom current activity. It is said that the density of suspended materials decreases with time, and it takes a year to one and a half years (Feely, 1976) for deposition to be completed.

The sedimentation processes of suspended particles in nepheloid layers involve the settling of individual particles, the coalescence of fine particles and their deposition by gravity, and the adhering of minute particles onto larger particles settling through the water column. According to Jacobs *et al.* (1973), the ratio of settling of individual particles to that of adhered particles or flows flocculated particles in the nepheloid layer at the depth of 5000 m in the North Atlantic are 84~96 % for the former and 1~12 % for the latter, of which bioclasts occupy 0.2~4 % of the amount. Based on this fact, Jacobs *et al.* considered that most of the particles from the nepheloid layer are sedimented as individual particles, whereas Feely (1976) preferred the formation of adhered particles as significant in sedimentation. As mechanisms for adhering, ionic adsorption (Hahn and Stamm, 1970) and bacterial activity (Sheldon *et al.*, 1967) are important.

Concerning the origin of fine-grained suspended particles in nepheloid layers, most of them were supplied either from terrigenous areas by turbidity currents or by suspended materials of not-yet-settled fine-grained particles and unconsolidated muddy sediments that were eroded and suspended by bottom currents (Jacobs, 1978). The nepheloid layer in submarine canyons deserves study to understand the transition process from turbidity currents to nepheloid layers (Beer and Gorsline, 1971; Drake and Gorsline, 1973; Barker, 1976; Kolla *et al.*, 1976).

As mentioned above, the nepheloid layer plays a significant role in the transportation and deposition of sediments on the continental rise and has a close relationship with bottom currents.

### *Benthic storms*

As to the origin of bottom currents, the importance of benthic storms has recently been emphasized in relation to the thermohaline circulation mentioned at the beginning of this section (Gardner and Sullivan, 1981; Hollister and McCave, 1984). The near-bottom currents to be called 'benthic storms' are related to erosion and deposition of bottom sediments in the ocean (Richardson *et al.*, 1981; Weatherly and Kelley, 1982). Benthic storms are augmented intermittent currents of the fast near-bottom flow, which show the velocity of 15~40 cm/sec at 10~50 m above the bottom, the duration of 1~20 days (mostly 3~5 days), and on the scale of 4~145 km in width. Although some link with ocean-surface storms (Gardner and Sullivan, 1981) and relatively high abyssal eddy kinetic energy (Schmitz, 1984) are considered, the cause of the benthic storms remains uncertain.

### 7.3.4 Black shale and the Oceanic Anoxic Events: From plate tectonics to plume theory

Deep-sea drillings (DSDP Leg 36 in 1974, Leg 40 in 1974–75, Leg 41 in 1975, Legs 43 and 44 in 1975, Leg 47 in 1950, Leg 50 in 1976, and Legs 71–75 in 1980) carried out in the Atlantic Ocean on the ocean floor at the early stages of sea-floor spreading have revealed black shale layers with 1~30% organic carbons in the Cretaceous sequence from 85 Ma to 115 Ma in age. Similar muddy sediments have also been found on the non-seismic ridges in the Indian Ocean (DSDP Legs 25, 26 and 27 in 1972) as well as in the Central Pacific (DSDP Leg 32 in 1973).

The Atlantic Ocean at the early stage of spreading was a gigantic landlocked sea, much like the present Black Sea, where there was a closed circulation of deep current systems and a limited supply of open-sea waters. A unique reducing environment in the Atlantic Ocean was recognized in the Angola Basin and the Cape Basin in the South Atlantic, where deep-sea sediments were cored by DSDP Leg 40 in 1975. Drilled cores from Site 364 in the Angola Basin and Site 361 in the Cape Basin clearly showed black shale in the Aptian of the Lower Cretaceous, and those from Site 364 also reported a second euxinic sapropelic shale in the late Albian to Coniacian-Santonian of the Upper Cretaceous. Furthermore, in the Angola Basin, 262 layers of black shale, all rich in organic carbon (19‰), have been reported from Site 530, DSDP Leg 75. These black shale layers abound in pyrite and plant debris, suggesting a provenance rich in vegetation. The basin waters associated with black shale were highly reducing and, since tests of calcareous microfossils were dissolved, no fossils are found.

Early Cretaceous oxygen-starved landlocked seas at the early stage of the spreading of the Atlantic Ocean are comparable with the modern Black Sea, as stated above. The Black Sea, called by the Ancient World the 'Euxine Sea', and hence the word euxinic, is characterized by reducing environments. DSDP Leg 42B drilled a total of three sites (Sites 379, 380 and 381) in the Black Sea and revealed a complete Pleistocene and perhaps Pliocene section, indicating a playa lake with only occasional marine incursions for most of the Pleistocene (Hsü, 1978).

The Cretaceous Period was a time of a warm greenhouse climate and luxuriant vegetation. In particular, angiosperms or flowering plants developed and flourished rapidly after the Late Jurassic (Sun *et al.*, 2001) and it is suggested that a quantity of organic matter of plant-origin far exceeding that of the present may have been transported by river flows and turbidity currents across the continental shelf to the ocean floor. In particular, a mass-production of calcareous nano-planktonic flora such as coccoliths over the deep-sea to the continental shelf seas led to a significant depletion of oxygen in sea waters. Furthermore, the decrease of land areas due to sea level rise related to climatic warming also promoted high productivity of planktonic flora. The rise of climatic temperatures caused decreases in the latitudinal temperature gradient, reduced the power of bottom currents owing to the increase in surface water temperatures, and brought about a significant reduction of dissolved oxygen in deep-sea waters. Furthermore, an oxygen-deficient ocean floor environment may have become more anoxic as organisms consumed oxygen. This is the reason why black shales were developed (Schlanger and Jenkyns, 1976). Such a rapid change of ocean environments as in the depletion of oxygen in global marine waters is called the Oceanic Anoxic Event (OAE) (Schlanger and Jenkyns, 1976).

The Cretaceous black shale problem or bituminous facies problem, as was demonstrated by deep-sea drilling in the Atlantic Ocean in 1974 to 1975, has been developed into the problem of oceanic anoxic events in the Cretaceous Period (Schlanger and Jenkyns, 1976; Jenkyns, 1980). H.C. Jenkyns (1976, 1980) further investigated the widely distributed Cretaceous black shale stratigraphy, which was deposited in deep-sea and shallow marine environments, and recognized three major occurrences of black and bituminous shales in the late Barremian–Aptian–Albian, at the Cenomanian–Turonian boundary, and in the Coniacian–Santonian. These intervals are considered to represent the duration of the Cretaceous Oceanic Anoxic Events (OAEs) (Fig. 7.24). At present, the chronology of the Cretaceous OAEs has been enhanced, revising them as follows: OAE1 for the Aptian–Albian, OAE2 (Bonarelli Event) for the Cenomanian–Turonian, OAE3 for the Coniacian–Santonian (Arthur *et al.*, 1990). The OAE1 is further divided into OAE1a for the earliest Aptian (Selli Event), OAE1b for the earliest Albian, OAE1c for the late Albian, and OAE1d for the latest Albian (Erbacher *et al.*, 1996). In Japan, the details of anoxic environments in the

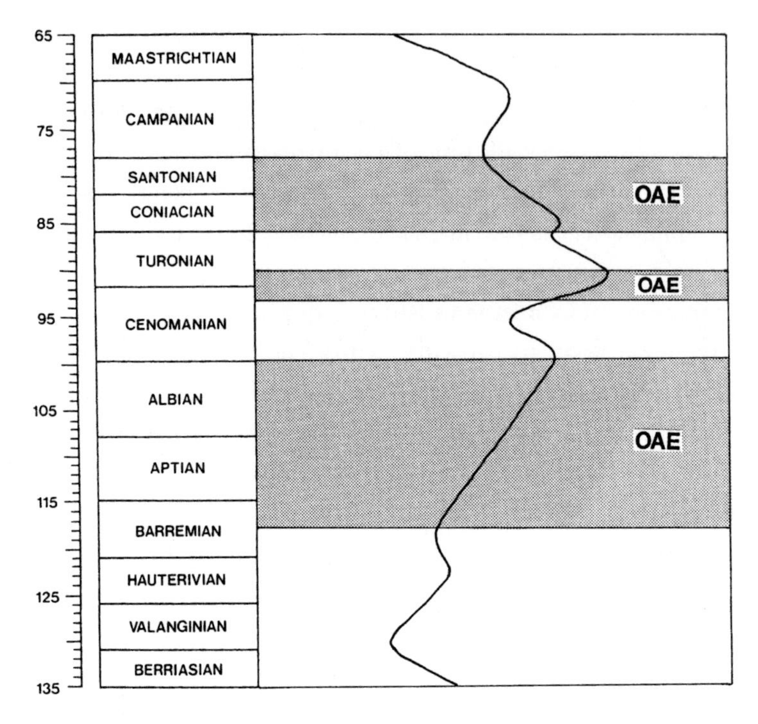


**Fig.** 7.24 Three major Oceanic Anoxic Events (OAE) (modified after Jenkyns, 1980). Relative transgressive curves are plotted against OAE. Transgression to the right, and regression to the left.

OAE2 interval (C/T boundary) have been determined using aspects of biological communities, sedimentary facies, trace fossils and isotope geochemistry (Hirano *et al.*, 1991, 2000; Hasegawa, 1995).

It seems that the Ocean Anoxic Events must contribute to the study of the important theme of plume tectonics (Maruyama, 1993), which may play a significant role as a key to the solution of the unique Cretaceous environmental problems in Earth history. Based on the theory of plume tectonics, the Larson Model has been proposed to explain the Cretaceous Earth environments that were established in response to rising superplumes in the Cretaceous, which in turn were responsible for intense volcanic activity, the enormous volume of world ocean-crust production mainly in the mid-Pacific in the late Barremian and early Aptian, longer than usual magnetic events, rises of carbon dioxide pressures, increased palaeotemperatures and eustatic sea level, global anoxic events, and world oil resources (Larson, 1991a, b) (Fig. 7.25). The Larson Model is explained as follows: the mid-Cretaceous was a time of greenhouse climate

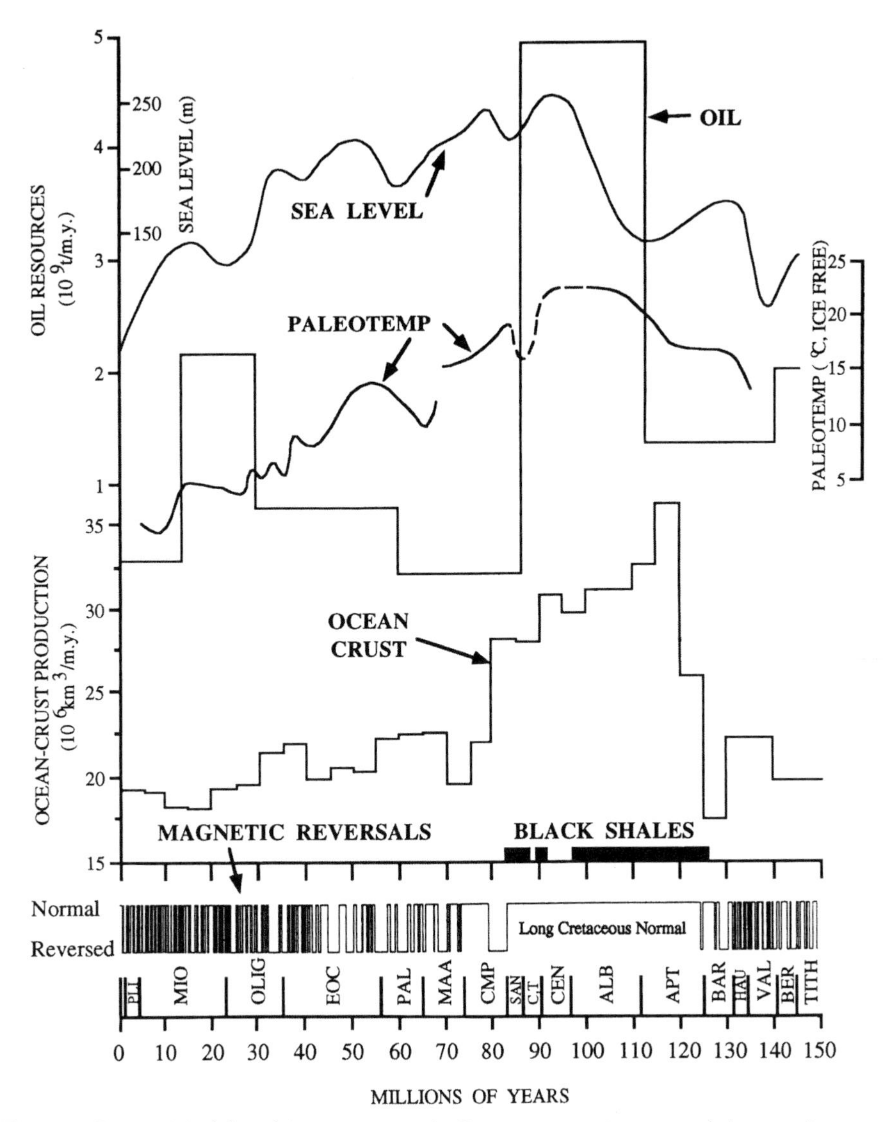


**Fig. 7.25** Larson Model explaining systematic Cretaceous environmental changes (Larson, 1991b). (*Reproduced by permission of the Geological Society of America.*)

characterized by reduced temperature gradients from the equator to the poles, absence of polar ice caps, and oceans at least 13° warmer than the present. Oceans were uplifted to extremely high stands of sea levels, and were more susceptible to development of oxygen deficits, expressed in various ways. Black shales were more widespread and 'anoxic events' were recorded by condensed oil-shale sequences in widely separated parts of the world. Conditions were particularly favourable for petroleum generation. The eruption of enormous volumes of basalt, mainly in the late Barremian and early Aptian in the mid-Pacific, may have played a major role in the initiation of the greenhouse effect by dramatically increasing $CO_2$ concentrations in the atmosphere (Larson *et al.*, 1992).

The theory of the OAE has been further applied to the Jurassic Period, the Permian/Triassic boundary and other Palaeozoic times (Isozaki, 1993, 1997). According to Isozaki, a superanoxia over 10 Ma long took place in the super-ocean Panthalassa surrounding the super-continent Pangaea at the time of the Permo-Triassic boundary about 250 Ma ago, which caused a great mass extinction.

As mentioned above, the continuing study of the problems of oceanic anoxic events following the discovery of the Cretaceous black shales by deep-sea drilling in the Atlantic Ocean will surely contribute much more to understanding Earth environments of the past and the future. In addition, new information of the computer tomography (CT) (Hager *et al.*, 1985) or geotomography of the Earth's interior has established the plume (or superplume) system in the lower mantle (670 km to 2900 km deep). It is suggested that superplume activity controls global tectonics including plate tectonics.

The theory of plate tectonic movement patterns was apparently settled in the 1980s (E. J. W. Jones, 1999), but new information from biostratigraphy (Tornaghi *et al.*, 1989) and of volcanism (Tarduno *et al.*, 1991) since 1989 have supported a new theory of plumes (Larson, 1991a, b).

# 8

# Sedimentology in Japan

The study of sediments and sedimentary rocks in Japan advanced with the development of geology. Geology had been introduced into Japan in the 1870s shortly after the Meiji Restoration of 1867, which led to the modernization of Japanese society. Although the history of sedimentology of Japan has been recently summarized elsewhere by Okada (2003), some new and significant facts are added in this chapter.

## 8.1 Geology and stratigraphy in their early stages

What follows is a brief history of the study of stratigraphy during the early stages of the geological investigation of Japan.

In 1867, immediately after the Meiji Restoration and following the Meiji Civil War, the Japanese central government invited European and American scientists to establish modern science and technology in Japan. The contributions of geologists and mining engineers were to be of major significance. Among them, Francisque Coignet (1835–1909) from France, Benjamin Smith Lyman (1835–1920) from the United States of America and Edmund Naumann (1854–1927) from Germany established the foundation of modern Japanese geology. Coignet, a French mining geologist, had been first invited early in 1867 by Nariakira Shimazu, a lord of the Satsuma regional government in Kagoshima and the most powerful governor in Japan before the Meiji Restoration, in order to survey and develop the mineral resources of the Kagoshima area of southern Kyushu. After the Meiji Restoration, Coignet's position was renewed by the Meiji central government and he worked to develop many mineral mines in Japan. He compiled for the first time the general characteristics of the geological structure of the Japanese Islands, paying special attention to the distribution of coal-bearing Tertiary strata in northern Kyushu (Imai, 1966).

Lyman, the American, had been educated by James Hall of the New York Geological Survey and the well known proposer of the geosynclinal concept (*see* Chapter 2.5). After being invited to Japan by the Hokkaido Development Agency in 1872, he made determined efforts not only in geological surveys in Hokkaido but also in the education of young geologists. His major achievements are represented by the geological sketch map of Hokkaido published in 1876 (Lyman *et al.,* 1876) and *A General Report on the Geology of Yesso* (Yesso or Yezo is the old name of Hokkaido) published in 1877 (Lyman, 1877). In these publications, he established the following seven formations in the geology of Hokkaido; the first lithostratigraphy established in Japan. His stratigraphy is shown here in descending order (the corresponding present-day stratigraphic divisions with their thicknesses are shown on the right) (Lyman *et al.*, 1876):

| | |
|---|---|
| **New Alluvium** | **[= Alluvium]: 100 feet** |
| **Old Alluvium** | **[= Pleistocene Series]: 100 feet** |
| **New Volcanic Rocks** | **[= Quaternary Volcanics]: 200 feet** |
| **Toshibets Group** | **[= Neogene System]: 3000 feet** |
| **(Middle Tertiary)** | |
| **Old Volcanic Rocks** | **[= Tertiary Volcanics]: 3000 feet** |
| **Chiefly Trachytic Rocks and Tufa** | |
| **Pebble Rocks of Same** | |
| **Horomui Group** | |
| **(Lower Tertiary ? Brown Coal Measures)** | **[= Palaeogene & Cretaceous]: 6500 feet** |
| **Kamoikotan Group** | **[= Kamuikotan Metamorphic Rocks]: 3000 feet** |
| **(Metamorphic without fossils)** | |

After his geological surveys in Hokkaido, Lyman gave all his efforts from 1876 to 1881 to surveying the oil fields of Japan. He attempted to apply his stratigraphy established in Hokkaido to the stratigraphy of the Niigata oil fields and also to the coal stratigraphy of Kyushu (Imai, 1966).

The German, Edmund Naumann, after learning German geology in its golden age at the University of Munich, worked in the geology section of the Bayern Mining Bureau. In 1875 he was invited as a professor to the Tokyo Kaisei School, which had been established by the central government of Japan to teach European science and technology. In 1877 this school was reformed as the Tokyo Imperial University and he was appointed as its first professor of geology (1877–1879). It was in this way that the systematic education of geology at university level was initiated in Japan. An assistant professor under him was Koreshiro Wada.

Naumann's significant achievements are represented not only by his establishment of the teaching of geology in Japan but also by his efforts in creating the Geological Survey of Japan in 1878 as a public organ to develop geological surveys in the Japanese Islands. One of his first students was Bunjiro Kotoh, who was later appointed as the first Japanese professor of geology.

It goes without saying that the foundation of the Geological Survey of Japan promoted Japanese geology and soon clarified the general features of the geology, stratigraphy and structure of the Japanese Islands. In particular, Naumann determined the geological and tectonic framework of the Japanese Islands in his book entitled *Uber den Bau und die Entstehung der Japanischen Inseln* (1885), in which he defined two major tectonic lines: the 'Fossa Magna', dividing the Japanese Islands into Northeast Japan and Southwest Japan, and 'the Median Tectonic Line' dividing Southwest Japan into the Inner Zone (on the Japan Sea side) and the Outer Zone (on the Pacific side).

It can be stated with confidence that these three invited foreign geologists established the early history of the development of Japanese geology. Soon after Naumann returned to Germany in 1885, the geological study of the Japanese Islands passed to Japanese geologists. Bunjiro Kotoh (geologist) and Toyokichi Harata (palaeontologist), both graduates of Tokyo Imperial University, were appointed as professors of that University, and Koreshiro Wada became the first Director of the Geological Survey of Japan. Thus, the geological study of Japan by Japanese geologists started.

In 1893 the Geological Society of Tokyo was founded, which was renamed the Geological Society of Japan in 1934. In 1912 a second Geology Department was opened this time at Tohoku Imperial University, the third was established at Kyoto Imperial University in 1921, a fourth at Hokkaido Imperial University in 1930, and the fifth at Kyushu Imperial University in 1939. Thus, the geological higher education system of Japan was established in the 1930s. As to noteworthy results, Hanjiro Imai (1921) developed the study of the Palaeogene stratigraphy of the Ishikari coal field in Hokkaido; Yoshiaki Ozawa (1923) revealed the Palaeozoic stratigraphy of the Akiyoshi Limestone in western Honshu; Takumi Nagao (1926–1928) established the Palaeogene stratigraphy in northern Kyushu; and Hisakatsu Yabe (1927) established the Cretaceous stratigraphy of the Japanese Islands. These studies indicate the rapid achievements in the stratigraphic study of sedimentary rocks in Japan. In this respect, it should be pointed out that Imai and Nagao already used many modern technical terms in their scientific descriptions.

## 8.2 Development of the study of sedimentary rocks and the significance of lithology

### 8.2.1 Development of the study of sedimentary rocks before 1940

Sedimentological studies of Japan had begun to appear before the international scientific community from around 1930. The representative works in this respect were the study of origin of petroleum by Jun-ichi Takahashi (1887–1959) (Fig. 8.1), a professor of Tohoku Imperial University (Takahashi, 1922) and his joint works with Tsugio Yagi on the origin of glauconites (Takahashi and Yagi, 1929a, b; Yagi, 1930; Takahashi, 1939). Takahashi and Yagi (1930) examined glauconite deposits, both ancient and recent, from Cretaceous, Tertiary and modern sediments in Japan, revealing five distinct modes of occurrence, namely: ovoidal glauconite; granular glauconite; pigmental glauconite with unclear outline gradually changing to matix; infilling glauconite filling cavities of fossils and fissures in minerals; and substituted glauconite replacing faecal pellets and fragments of volcanic glasses, minerals, and fossils. They presented the chemical compositions of each of them.

**Fig. 8.1** Jun-ichi Takahashi (courtesy Mr. H. Takahashi).

In particular, it was noted that glauconites being formed in the recent sediments of Aomori Bay, northernmost Honshu, Japan, were derived from faecal pellets (Takahashi and Yagi, 1929a,b). This study of glauconites was internationally highly acclaimed (Twenhofel, 1932, 1961).

In addition, Takahashi contributed much to the study of the origin of petroleum (oil and natural gas), in which he attached great importance to a material called kerogen (Takahashi, 1922). He also studied sedimentary cycles in the oil-bearing Miocene Oga Series in northern Honshu, Japan (Takahashi, 1940). As these studies indicate, Takahashi was indeed a pioneer of sedimentology in Japan. The research style of Takahashi and Yagi was exhibited in their publications containing such modern terms as sedimentary province, sedimentary environment, sedimentary facies, sedimentary cycles and so on, as early as in 1929 to 1935. What is more, Yagi (1897–1951) (Fig. 8.2) introduced a new term '*taiseki-gaku*' ('sedimentology' in English) (Yagi, 1929, 1933, 1935). The introduction of this term was its first use in the world, though the use of the new term was not accepted by other geologists in Japan for another 20 years (*see* below).

**Fig. 8.2** Tsugio Yagi (courtesy Prof. T. Kato).

On the other hand, T. Nagao (1926–1928) carried out a systematic study of the coal-bearing Palaeogene formations of Kyushu, using both old and modern sedimentological terms. However, T. Matsumoto (1936, 1938) preferred to use only modern terms. Through such publications, some sedimentological words gradually became familiar for the description of sediments and sedimentary rocks.

As a representative work on sedimentary rocks at this time, the stratigraphic study of Palaeogene coal fields in northern Kyushu by H. Matsushita (1949) was important. Criticizing the conclusion of Nagao (1926–1928) that the whole of the Palaeogene sediments of northern Kyushu were deposited in a single basin and that variations in lithofacies at different places were simply due to different sedimentary environments within that basin, Matsushita argued that the sedimentary basin of the Kyushu Palaeogene was not a single entity, but had been divided into a number of half-graben-like small basins produced by tensional faults in the basement under the control of regional tectonic movements. Based on further detailed studies of the nature of sediments in each small basin and their correlations, he subdivided the Palaeogene of northern Kyushu into the following five stages: Amakusa Stage (Palaeocene–Eocene), Ariake Stage (early Eocene), Nogata Stage (late Eocene), Otsuji Stage (latest Eocene–earliest Oligocene), and Chikushi Stage (Oligocene–Miocene), and presented the palaeogeography at each of thirteen selected times (Fig. 8.3). This study was a significant contribution to understanding the development of Palaeogene sedimentary basins in northern Kyushu. The procedure and description in the study became a benchmark for modern studies of sedimentary rocks in Japan.

Other important studies of sedimentary rocks were published by Matsumoto (1940). He made a petrological study of sandstones from the uppermost Cretaceous Ryugase Formation (correlated with the Hakobuchi Group of Hokkaido, Japan) in southern Sakhalin Island (Siberia, Russia), and showed that these sandstones are characterized by large amounts of acidic to intermediate volcanic materials. He further emphasized the importance of volcanism in the development of the Mesozoic Yezo–Sakhalin geosyncline. It is interesting to note that this petrological feature of the Ryugase Formation is similar to that of the sandstones of the Hakobuchi Group in Hokkaido (Okada, 1974).

In addition, in 1937 a noteworthy book entitled *Limnology* was published by Shinkichi Yoshimura, the founder of modern limnology in Japan. It was later revised in 1976. Yoshimura described in detail the nature of lake sediments, strengthening their importance as indicators of environmental changes.

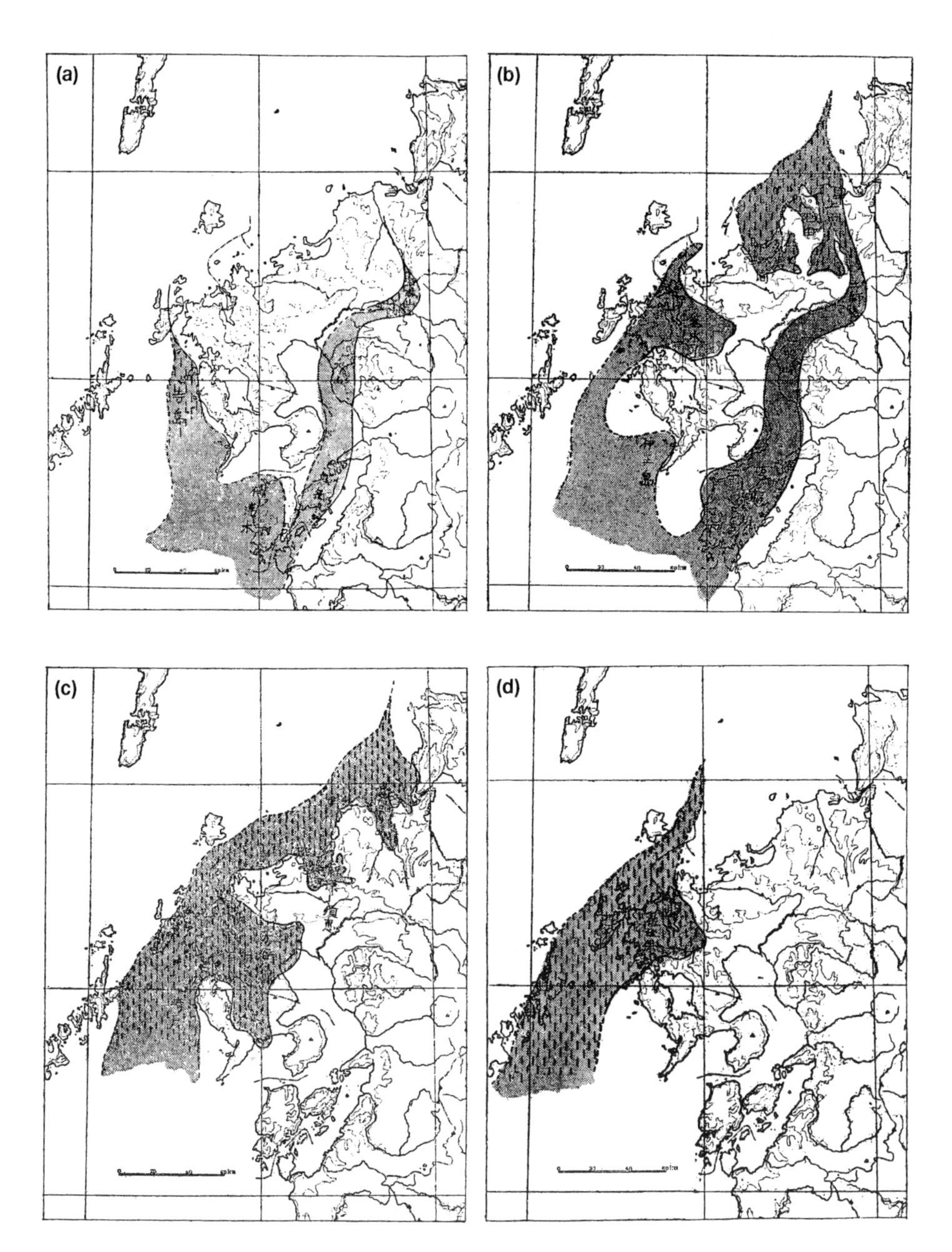


**Fig. 8.3** Palaeogeography of the Palaeogene coal-bearing successions in North Kyushu (Matsushita, 1949). (a) Ariake Stage (early Eocene), (b) Nogata Stage (late Eocene), (c) Otsuji Stage (Oligocene), (d) Late Chikushi Stage (Miocene). Dark parts indicate the sea area, and dark parts with dotted lines show the freshwater area.

### 8.2.2 The lithology concept in Japan

The publication in 1931 of a textbook entitled *Chiso-gaku* ('Lithology' in English) by Hanjiro Imai (1884–1938) (Fig. 8.4) is of special significance. Imai stated in his preface that the purpose of the book was to encourage the systematic study of sediments and sedimentary rocks in Japan, because no book was available in Japanese treating sediments and sedimentary rocks was available even though the study of the earth sciences had advanced remarkably. He further said that 'lithology is the science which studies all the aspects of sediments and sedimentary rocks'. This concept of 'lithology' is akin to the definition of 'sedimentology' given first by Hakon Wadell (1932b) (Okada, 1998a). Imai's book was indeed a masterpiece reflecting the international awareness of Japanese geologists at that time. In his book Imai translated many English terms into Japanese. The usage of the term *Chiso-gaku* (lithology) as the title of the book was very innovative even by today's standards. In addition, he introduced other sedimentological words into his earlier papers (for examples, *see* Imai, 1921).

Later, Masao Minato (1953) explained the term 'lithology' as 'dealing with fundamental problems concerning strata' in his book with the same title *Chiso-gaku.* In the second edition of his book (Minato, 1973, p. 9), he noted

**Fig. 8.4** Hanjiro Imai (Yamaguchi, 1939).

that '*Chiso-gaku* makes it a basic principle in understanding evolutionary developments of the sediments and sedimentary rocks of the Earth's crust'. By this expression, he seems to have made efforts to promote *Chiso-gaku* to a much broader scientific level. This concept of the new discipline correlates with the 'stratigraphic geology' of Tatsuro Matsumoto (1949).

During the 1930s the concept of 'lithology' was widely developed in Russia (Strakhov, 1970). It is uncertain, however, if the Russian School influenced the establishment of the concept of *chiso-gaku* in Japan. Nevertheless, since the early history of Japanese geology had been influenced by German geology, it can be suggested that there may well have been an indirect influence on the Japanese use of lithology from Johannes Walther's facies concept (*see* Chapter 6).

Unfortunately, however, neither Imai (1931) nor Minato (1953) had given an English translation to *chiso-gaku*. Their concept of this term can with certainty be compared to the 'lithology' of the Russian School (*refer* to Chapters 5.4 and 6.1.2). The author has, therefore, applied the term 'lithology' to the Japanese *chiso-gaku* as the most appropriate English translation. Previously, however, even in Japan the English term 'lithology' had been used for describing the physical characters of rocks (Association for Geological Collaboration, 1970, p. 231), and thus it should be emphasized that the present term 'lithology' has been given a wider meaning than before (G. V. Middleton, 1978 and his personal correspondence). In this respect, it is interesting to note that the term 'lithology' has been used in the United States of America to describe either the physical character of any or all (i.e. igneous, sedimentary or metamorphic) rocks or their petrography (Bates and Jackson, 1987).

## 8.3 Establishment of sedimentology in Japan

The study of sediments and sedimentary rocks in Japan before the 1950s was summarised in Matsumoto's (1949) book *Assignments of Geological History of Japan* and in Minato's (1953) *Chiso-gaku*. Based on the concept of stratigraphic geology, Matsumoto (1949) inquired into the geological history of Japan. On the other hand, Minato (1953) grasped the general features of strata on the basis of the lithology concept. Both of these thoughts predicted the coming of sedimentology in Japan.

The scientific thinking close to modern sedimentology was first introduced in Japan by Hanjiro Imai (1931), as described in the foregoing section. However, the term *taiseki* ('sedimentary, sedimentation' in English) had never been generally used in Japan at that time, and instead the word *chinseki* (comparable to 'settling' in English) was more common. Thus, at the time of Matajiro Yokoyama, a pioneer of Japanese geology, all sedimentary rocks were called *chinseki-gan* (*gan* is rock in English) (Yokoyama, 1914). Even in the textbook of *Petrographic Geology* by Denzo Sato (1924, 1940), such words as aqueous rock, clastic deposit, organic deposit and chemical deposit were used.

Incidentally, it is noteworthy that Takumi Nagao, Jun-ichi Takahashi and Tsugio Yagi, all sedimentary researchers at Tohoku University, as early as the 1920s had introduced the terms sedimentary facies, sedimentary environments, sedimentary cycles and so on into their descriptions, all of which are still in common use. At first, Takahashi and Yagi tried to use the new Japanese term *taisekigan-gaku* ('science of sedimentary rocks' in English) (Yagi, 1929, 1930; Takahashi and Yagi, 1929a). Eventually Yagi used *taiseki-gaku* ('sedimentology' in English) in his paper in 1929 (Yagi, 1929). Further, in 1933, he gave the term *taiseki-gaku* to the title of his paper (Yagi, 1933). It was almost contemporary with the introduction of the term sedimentology by Hakon Wadell in 1932 (*see* Chapter 6.3). Strictly speaking, Yagi's usage of the term *taiseki-gaku* was earlier than that of 'sedimentology' by H. Wadell.

In spite of the early work of Yagi possibly, the usage of the term *taiseki-gaku* (sedimentology) to denote the study of sediments and sedimentary rocks was only widely accepted after World War II. The Japanese dictionary of geology published in 1935 (Watanabe, 1935) still used the word *chinseki* more frequently than *taiseki*.

The situation, however, changed remarkably when Takao Sakamoto (1890–1980) (Fig. 8.5) became the professor of sedimentology at the Department of Geology, University of Tokyo, in 1949. He studied sedimentology under Prof. W. H. Twenhofel at the University of Wisconsin in 1927. At that time he studied Precambrian iron formation near Lake Superior under the guidance of Prof. Leith. This experience led to his study on the origin of Precambrian banded iron formations in Northeast China (Sakamoto, 1950). Incidentally, Shoji Ijiri (1950) gave sedimentology a prime scientific position in geology as a modern fundamental science in his paper. Further it may be noted that marine geology in Japan began to develop quickly at this time, as represented by the works on submarine sediments by Risaburo Tayama (1950) and Hiroshi Niino (1950).

As stated above, the early 1950s was the time of the establishment of sedimentology in Japan. Research by stratigraphers and palaeontologists was

**Fig. 8.5** Takao Sakamoto (Nasu, 1981).

remarkably active at that time, especially on the problems of sedimentary facies, sedimentation and sedimentary environments.

At the same time, sedimentology was also being established internationally (Okada, 1998a). Its general acceptance started around the early 1960s. In this respect, therefore, it may be said that sedimentology in Japan came into existence about ten years ahead of the international trend.

## 8.4 Development of sedimentology in Japan

The present arrangement of sedimentological studies in Japan started as early as the beginning of the 1950s. Three distinct stages can be identified: the first was the foundation of the Sedimentological Subcommittee of the National Committee of Geology in 1951, the second was the establishment of the Sedimentological Association in 1957 and the third was the establishment of marine science institutes after 1962.

### 8.4.1 Foundation of the Sedimentology Subcommittee of the Science Council of Japan

The foundation of the Sedimentology Subcommittee of the National Committee of Geology, one of the major committees of the Science Council of Japan, occurred in 1951 (Nasu, 1961). The Subcommittee (first chairman: Prof. Takao Sakamoto, University of Tokyo) wished to direct researchers of sediments and sedimentary rocks. For this purpose, it attempted to organize sedimentary researchers of modern sediments into collaborative research projects with governmental financial support, for example, on 'The collaborative study of sedimentology'. A serial newsletter for this project, *Renkon* (*Ripple Mark*) was published until its fifth issue. The title was, however, changed to *Taiseki-gaku Kenkyu* (*Study of Sedimentology*) from the sixth issue in 1952. Afterwards, new collaborative research projects on the 'Ecological and Sedimentological Studies in Matsukawa-ura Bay, Sohma City, Fukushima Prefecture' (1953–1954; Leader: Prof. Kiyoshi Asano, Tohoku University), 'Collaborative Study of Recent Submarine Sediments in Tokyo Bay' (1955; Leader: Prof. Hiroshi Niino, Tokyo Fisheries College), and 'Collaborative Study of Recent Submarine Sediments in Tokyo Bay and Adjacent Seas' (1958; Leader: Prof. Takao Sakamoto, University of Tokyo) were organized.

Prof. Kiyoshi Asano, the project leader of the study of Matsukawa-ura Bay, which was the first comprehensive marine sedimentological study in Japan, described the purpose of this study as follows: 'While sedimentological and ecological studies around the Japanese Islands had hitherto been carried out by individual workers, research methods and interpretations were limited and the results became naturally biased whilst we would like to observe fossil-bearing strata with respect to sedimentology.' Therefore, Asano (1955) organized this project based on such variable disciplines as geomorphology (by Hisao Nakagawa), geology and geological structure (by Nobu Kitamura *et al.*), sedimentology (mainly on ripple marks) (by Hideo Mii), foraminifers (by Yokichi Takayanagi), molluscs (by Tamio Kotaka *et al.*), crabs (by Rikizo Imaizumi), submarine environments (by Hideo Mii), and benthonics and their environments (by S. Noda). The participants in this project were able to provide significant and fundamental data on recent submarine sediments in a designated bay – results which are still highly appreciated.

As stated above, the strategy of the Sedimentology Subcommittee in organizing and supporting collaborative studies was very effective. On its successful conclusion, the Subcommittee's activities were terminated in 1959 and instead a new Marine Geology Subcommittee, chaired by Prof. Hiroshi Asano, started (Nasu, 1961). Nevertheless, there was an eager desire by sedimentologists to re-establish the Sedimentology Subcommittee, and this was accepted by the Science Council of Japan in 1982. Since then, the Sedimentology Subcommittee has been active not only in supporting the Japanese community of sedimentologists but also in working as a liaison group to the International Association of Sedimentologists (IAS) (Okada, 1998b).

### 8.4.2 Establishment of the Sedimentological Association

The second line that defined the trend of sedimentology in Japan was the establishment of the Sedimentological Association as an organization for supporting activities of sedimentologists. The sedimentologists' group of the collaborative research projects organized in 1953 to 1955 established the Sedimentological Association in 1957, the secretariat office of which was located at the Natural Resources Research Institute (presently, National Science Museum Annex at Shinjuku, Tokyo). It was the first official organization of Japanese sedimentologists and it started to publish a journal *Taiseki-gaku Kenkyu* (*Sedimentological Research*). The first issue of the new journal was published on 1 September 1957, and it continued until the 22nd/23rd issue in 1960. All the issues were only of about ten pages in B5 size, but contained such items as summaries of recent works, international news, new literature and so on, which were quite useful to sedimentary researchers. Sadly, however, the Sedimentological Association gave way to the Marine Geology Association in 1962. Its first president was Prof. Hiroshi Niino, professor at Tokyo Fisheries College and he was succeeded in 1965 by Prof. Noriyuki Nasu, professor of the University of Tokyo. The Marine Geology Association began to publish the journal *Marine Geology* in 1962, but this was discontinued in 1972 with the last issue of No. 2 of Vol. 8.

In these circumstances, the defunct Sedimentological Association was restarted, thanks to a great effort by Prof. Rikii Shoji, professor of Tohoku University, in order to encourage further developments in sedimentology. It was reorganized under the name of *Taiseki-gaku Renrakukai* (Liaison Society of Sedimentology), and Prof. Kagetaka Watanabe, professor of Tokyo University of Education (presently, Tsukuba University) was nominated as the president. The secretariat office for it was set up at the Geological Survey of Japan, located at that time in Kawasaki City south of Tokyo. Dr. Atsuyuki Mizuno served as its secretary. The present organization was renamed as the Sedimentological Society of Japan in 1971 and has retrieved its original status. Its official journal was first called *The Journal of the Liaison Society of Sedimentology* (1969–1971), then changed to *The Journal of Sedimentological Research* (1972–1993) and finally became *The Journal of the Sedimentological Society of Japan* (after 1994). Each issue of the present journal comprises about 90 pages on average with a modified A4 size. Not only has its cover design been revamped but also the contents have been remarkably improved, and now it accepts papers from overseas. Furthermore, the Sedimentological Society of Japan published an *Encyclopedia of Sedimentology*

(Sedimentological Society of Japan, 1998) in Japanese, which has been highly appreciated internationally.

The reason why the organization of Japanese sedimentology was so unstable is not clear, but the creation of the Sedimentological Society of Japan, established in 2002, seems to have created a more satisfactory arrangement for sedimentologists in Japan.

### 8.4.3 The establishment of marine science institutes and their effects on sedimentology

The third theme in the rapid development of sedimentology in Japan owes much to the establishment of Japanese marine science institutes. In 1962 the Ocean Research Institute was founded by the University of Tokyo, and in 1974 a Department of Marine Geology was established in the Geological Survey of Japan; both have provided sedimentologists with cooperative facilities for their work. In particular, the submarine sedimentation section under Prof. Noriyuki Nasu of the Ocean Research Institute, University of Tokyo, has given great support to sedimentologists by providing the research vessels *Tansei-maru* (260 t) and *Hakuho-maru* (3,000 t). Sedimentologists in Japan have, consequently, obtained great many opportunities in new research areas as well as in broader scientific themes.

In addition, the Ocean Research Institute, University of Tokyo, had played an important role in the Japanese participation in the International Phase of Ocean Drilling (IPOD) since 1975. Further, the Japan Marine Science and Technology Centre (JAMSTEC) was established in 1971 as a frontier institute in deep-sea research. It has been provided with a 2000 m depth capability submersible *Shinkai 2000* (launched in 1981) and a 6500 m depth capability submersible *Shinkai 6500* (launched in 1990). These vessels have been indispensable for sedimentological surveys in trenches and other deep sea regions around the Japanese Islands.

### 8.4.4 Long-range plan in earth sciences: 'Experimental Sedimentology Institute'

A special event in the 1960s was a request from the 6th Science Council of Japan to the Committee of Geology in 1961 to discuss how to improve the long-range science system of Japan. Thus, the Committee of Geology founded a Sub-committee of the Long-range Science System and began to discuss, under the chairmanship of Prof. Tatsuro Matsumoto, professor of Kyushu University, the construction of a National Institute of Experimental Sedimentology. As a result, two plans for constructing the Institute of Experimental Sedimentology were proposed (Omori, 1968). One of the proposals said that the Institute should be composed of the following eight divisions: dynamic physical research, mineralogical research, inorganic chemical research, organic chemical research, biological research, volcanic research, lithological research, and mathematical and statistical research.

Unfortunately, this epoch-making proposal was never realized, but it did effectively encourage the activities of sedimentologists in Japan.

### 8.4.5 Development of sedimentological studies

As the framework of Japanese sedimentology has developed, the number of sedimentologists has increased remarkably and their works have rapidly improved in quantity and quality (Shoji, 1966). In this section, some representative studies grouped into ten-year periods are illustrated.

#### *1950–1960*

It was the individual pioneers of the 1950s who started a major developmental trend in Japanese sedimentology. Their contributions to the international community of sedimentology gradually became more significant, as represented by the studies of dynamic sedimentary processes by Haruhiko Kimura (1953 1956) and Rikii Shoji (1955), petrologic studies of Mesozoic sandstones in Middle Kyushu by Koji Fujii (1956), heavy mineral studies of Palaeogene sandstones in Hokkaido by Azuma Iijima (1959), studies of the optical characters and provenances of clastic feldspars by Shinjiro Mizutani (1959), petrological studies of Palaeozoic sandstones in the Maizuru Belt in Kyoto by Tsunemasa Shiki (1959) and sedimentological studies of Tertiary sedimentary rocks in the Chichibu Basin north of Tokyo (Juzo Arai, 1960).

In this period, Jiro Makiyama, a professor of Kyoto University, developed a unique hypothesis on the relationship between tectonic movements and sedimentology thus supporting the work of Rich (1951). Makiyama (1956) divided the sedimentary environments of geosynclines into clinothem for a delta-front slope, fondothem for an off-shore deep floor, and undathem for the upper part of a delta-front slope, and discussed the development of sedimentary basins.

In addition, there was a useful textbook entitled *Petrography [IV]* on sedimentary petrography by Nobuo Aoyama (Aoyama, 1958), which became familiar among students at a time of few Japanese publications on geological sciences.

### *1960–1970*

In this period many systematic studies were carried out both on clastic and non-clastic sedimentary rocks. In particular, diverse studies of Mesozoic to Cenozoic flysch facies were actively practiced. Further in 1967, a very successful symposium on 'Some Problems in Sedimentology' held at Nagoya University made a major impact on many geologists and sedimentologists.

The following are significant works on some selected topics of sedimentology in this decade:

**Sedimentary geology of flysch sediments**: Studies of geosynclinal sediments by the methods of sedimentary petrology and palaeocurrent analysis were most active during this period. Three dimensional traces of flysch-type alternations of sandstone and mudstone by Hirayama and Suzuki (1968), palaeocurrent analyses of the Cretaceous Yezo Supergroup flysch in Hokkaido by Tanaka (1963, 1970), lithofacies analyses of the Cretaceous Izumi Group in Kii Peninsula by Tanaka (1965), basin analysis of the Cretaceous Ohnogawa Basin in Middle Kyushu by Teraoka (1970), studies of sedimentary structures and trace fossils in the Cretaceous Shimanto Supergroup by Katto (for example, 1960, 1961), palaeocurrent analyses of the Cretaceous to Palaeogene Shimanto by Harata (1965), bed-thickness analysis of turbidite sandstones in the Shimanto by Kimura (1966), and the discovery of orthoquartzite gravels from the Shimanto by Tokuoka (1970) were most important.

**Sedimentary petrology of clastic rocks**: Studies by Ohara (1961), Shiki and Mizutani (1965), Okada (1968a) and Mikami (1969) were important. Hayashi (1966) made a unique study of the nature and genesis of clastic dykes.

**Sedimentary geology and petrology of carbonates**: Studies of the Carboniferous to Permian Akiyoshi Limestone in West Honshu by Ota (1968), of the Permian Sakamotozawa Limestone in North Honshu by Mikami (1969) and of facies analysis and sedimentary environments of the Permian to Triassic carbonates in the Sanpozan Belt, Kyushu, by Kanmera (1969) were of high quality.

**Diagenesis**: Studies of the diagenesis of siliceous minerals and sediments by Mizutani (1966, 1970) and of pyroclastic sediments by Iijima and Utada (1966) and Utada (1965) were significant.

**Sedimentology of coal- and oil-bearing strata**: In addition to the book *Coal Geology* by Katsuhiko Sakakura (1964), palaeocurrent analysis of the Palaeogene coal-bearing strata in northern Kyushu by Nagahama (1965), the origin of sedimentary cycles in coal-bearing strata by Shoji (1960), and the geological and sedimentological features of oil-bearing strata by Ikebe (1962), Kinoshita (1962) and Aoyagi *et al.* (1970) were important studies.

**Analysis of submarine environments**: There were significant contributions by Mutsuo Hattori (1967) on Sendai Bay, Northeast Honshu, and by Hiromi Mitsushio (1967) on shallow seas off the coast of northern Kyushu. In addition, an introductory book *Guide for the Studies of Waters and Sediments* (in Japanese) by Atsuyuki Mizuno (1968) for the study of submarine sediments was very helpful for beginners.

### *1970–1980*

This was the period when the establishment of the Ocean Research Institute at the University of Tokyo, Department of Marine Geology at the Geological Survey of Japan and the Japan Marine Science and Technology Centre began to develop systematic marine research in Japan (*see* Chapter 8 Section 4.3) and sedimentologists rapidly contributed much to the sedimentological analyses of submarine sediments and to the study of submarine environments based on the enhanced resolving capability of the chronology of marine sediments in collaboration with palaeontologists (for example, Mizuno *et al.*, 1970; Kamada, 1979). Furthermore, by participating in such international projects as the International Phase of Ocean Drilling (IPOD) and WESTPAC Project by Western Pacific Subcommission of UNESCO/IOC, Japanese sedimentologists had many more opportunities to develop their activities.

Regarding on-land sedimentary geology, sedimentological studies of flysch-like sedimentary rocks in the Shimanto Belt in Southwest Japan were still active, as shown by Tateishi (1978) and Teraoka (1979). At this time, the new accretionary model for the arc–trench system known as the Seely Model (Seely *et al.*, 1974) was proposed, and the study of the Shimanto Belt as the trench-accretionary system developed remarkably mainly on the basis of radiolarian biostratigraphy established by the Deep Sea Drilling Project (DSDP) and newly understood tectonic and stratigraphic features of accreted sediments (Ingle *et al.*, 1975), as is well shown by Okada (1971c), Kanmera (1976, 1977), Sakai (1978) and Taira and Tashiro (1980).

In addition, sedimentological studies of the Upper Cretaceous Nemuro Group in Hokkaido were carried out by Kiminami (1975, 1975–1976). In particular, Hirayama and Nakajima (1977) successfully traced individual turbidite beds over a distance of 38 km in the Miocene to Pliocene flysch-like deposits in a forearc basin on Boso Peninsula, southeast of Tokyo. Analyses of three-dimensional features of sandstone-rich flysch facies by Tokuhashi (1979) and disturbed sedimentary structures by Yamauchi (1977) were additional important studies.

Furthermore, studies of chert (Saito, 1977; Iijima, 1978; Saito and Imoto, 1978) and coal sedimentology (Okami, 1973; Miki, 1977) became more active. A new automatic analyser of grain size was also developed by Niitsuma (1971).

Further worthy of mention was the important advance in mathematical sedimentology as in the study of stochastic analyses of sedimentary environments by Mizutani and Hattori (1972) and Mizutani (2000), as well as in Markov chain analyses of sedimentary processes in the Quaternary Ryukyu Limestone (Minoura, 1979). In addition, studies of weathering processes of granites and their weathering speed by Kimiya (1975a, b) were noteworthy.

Early in this period, the first textbooks on sedimentology and sedimentary petrology were published by R. Shoji (1971a, b), as was also the textbook *Science of Oil Resources* by K. Kinoshita (1973) on the generation, migration and accumulation of oil. Hirayama *et al.* (1967, 1968, 1971) completed the translation of N. M. Strakhov's (1962) textbook *Principles of Lithogenesis* into Japanese. All indicate the growing maturity of Japanese sedimentology.

### *1980–1990*

In this period sedimentological studies broadened. In addition some important publications appeared which have served to expand the basis of Japanese sedimentology. They were the books *Materials and Environments of the Earth's Surface* by Kanmera *et al.* (1979), *Procedures of Study of Sediments* by Research Group on Clastic Sediments (1983) and *Sedimentary Rocks of Japan* by Mizutani *et al.* (1987), all in Japanese. Major achievements in sedimentology in this period were as follows:

**Lithofacies analysis**: Lithofacies analyses based on directional data of sedimentary structures, proximality indices of turbidites and such sedimentary patterns as fining-upward sedimentation, deep-sea fan models and so on, all established in the 1960s to 1970s period, were actively applied to sedimentological studies (Hisatomi, 1984; Soh, 1985; Ito, 1985; Takizawa, 1985; Maejima, 1986; Hoyanagi, 1989; Tanaka, 1989; Takano, 1990).

**Study of muddy chaotic deposits**: Studies of olistostrome, slump deposits and mélange characteristic of orogenic-belt deposits, to which K. Kanmera (1977) called attention, progressed rapidly, as represented by Taira *et al.* (1982), Okada (1983), T. Sakai *et al.* (1987), H. Sakai (1988a, b, c) and Wakita (1988a, b).

**Study of formation processes of the accretionary prism**: Clear tectonic models of trench subduction and formational processes of the accretionary prism in the Nankai Trough exemplified by Taira (1981) and Kagami *et al.* (1983) made rapid progress in the sedimentological studies of island-arc orogenic belts.

Another significant contribution was the great success in the study of trench sedimentation by the French–Japanese KAIKO Project (*Kaiko* is trench in Japanese) using a new French research submersible *Nautile* in 1984–1985 (Le Pichon *et al.*, 1987a). In particular, discoveries of deep-sea channels meandering in the axis of the Nankai and Sagami Troughs (Le Pichon *et al.*, 1987b; Okada, 1988) and of a subduction process at the Daiichi Kashima Seamount (KAIKO I Research Group, 1985, 1986; KAIKO II Research Group, 1987; Cadet *et al.*, 1987) were really important.

**Study of sedimentary structures**: There were original experimental studies of liquefaction and the deformation of sediments (Tsuji and Miyata, 1987) and of hydraulic behaviours of depositional particles (Miyata, 1989).

**Study of sedimentary petrology:** Petrological studies of sandstones were carried out by Miyamoto (1980) and others. Furthermore it is important to mention that studies of the provenance and origin of clastics based on the chemical compositions of contained stable minerals were initiated as represented by the studies of clastic garnets by Adachi (1985), Takeuchi (1986) and Musashino and Kasahara (1986), which have encouraged further developments of such research.

**Study of diagenesis**: There were important studies such as the maturation process of kerogen materials by Ujiie (1981), Suzuki and Taguchi (1982) and Taguchi (1983) and the diagenesis of carbon-rich sediments by Aihara (1983). In particular, the study of diagenetic changes of organic materials in sediments contributed much to the understanding of the burial and thermal processes of sediments. Additionally, Matsumoto (1987a, b) paid attention to the importance of the study of gas hydrates.

**Study of sea-floor environments**: Ohshima *et al.* (1982) studied the sedimentation history of submarine sediments around the Tsushima and Goto Islands off northern Kyushu. Takayanagi *et al.* (1985) carried out collaborative studies of Late Quaternary palaeoenvironments from cored sediments drilled off Kashima, north of Tokyo, paying special attention to lithofacies, oxygen-isotope stratigraphy, microfossil communities and mineral compositions. Konishi *et al.* (1985) also made an interesting study of the submarine palaeoenvironments of the Okinawa seas using reef-drilled cores.

**Study of trace fossils**: Studies of trace fossils in Japan were initiated by Ichiro Hayasaka in the 1930s to 1960s, and developed by Jiro Katto and Tsutomu Utashiro in the 1950s to 1980s and Koji Noda (for example, Noda, 1994) in the 1970s to 1990s. In this final stage, Kotake (1989) made a unique study of the occurrence of *Zoophycos* and its genesis based on lithology, and since then the study of trace fossils in Japan has reached a new stage of development.

**Sedimentology of siliceous and carbonate rocks**: Studies of the geochemical features of radiolarian bedded chert (Matsumoto, 1982), of bedded cherts in the accretionary prism (Yamamoto, 1983) and in the Pacific Ocean (Yamamoto, 1988, 1989), and of pelagic sediments (Sano, 1988a, b, 1989) were important.

**Application of mathematical geology to sedimentology**: Newly developed methods of mathematical geology by using computer programmes and systems were actively applied to sedimentology in the 1970s (Nishiwaki and Yamamoto, 1981).

### *1990–2000*

Japanese sedimentology reached the stage of widespread acceptance measured not only in its breadth of research interests but also by the rapid increase in young sedimentologists at work. In particular, a well-balanced development between field observations, experimental studies and theoretical works has characterized this stage. Incidentally, studies of sequence stratigraphy were developed, keeping well in step with the international trend.

**Petrologic study of sandstones and analyses of their provenance**: In addition to the geochemical study of clastic garnets in sandstones developed in the 1990s (Takeuchi, 2000), the study of the provenance of clastic chrome spinels in sandstones has become active (for example, Hisada and Arai, 1993; Nanayama and Nakagawa, 1995; Suzuki *et al.*, 1997). It is noteworthy that the study of source rock compositions and the original tectonic environments on the basis of chemical compositions of sandstones has become significantly important (Kumon *et al.*, 2000). One of the representative achievements of this study is the Basicity Index (BI index) proposed for discriminating detailed provenances and tectonic environments of arc-volcanism-origin sandstones (Kiminami *et al.*, 1992, 2000; Kumon and Kiminami, 1994). The BI diagram is based on the relationship between $Al_2O_3/SiO_2$ and the BI index [(FeO + MgO) / ($SiO_2$ + $K_2O$ + $Na_2O$), which divides the tectonic environments of the arc into IIA (immature island arc), EIA (evolved island arc), CA (continental arc) and DA (dissected arc) (*see* Fig. 4.7).

**Actualistic sedimentation**: Excellent observations of wave dynamics and formation processes of ripple marks were carried out in the Ariake Inland Sea, northern Kyushu (Makino, 1994). Suzuki (1994) made detailed observations of flood deposits in modern rivers and pointed out the importance of reverse grading structures as a flood depositional facies.

**Event sedimentation**: As examples of event depositions taking place in quite a short time, seismite, tsunamiite and tempestite have been studied. In addition to the study of storm-related sediments by Chijiwa (1985), Shiki contributed much to the development of the study of event sedimentations (Shiki, 1993; Shiki and Yamazaki, 1996; Shiki *et al.*, 1996). Minoura and his co-workers showed criteria of tsunami deposits and their correlation with historic tsunami records (Minoura and Nakaya, 1991; Minoura and Nakata, 1994; Minoura *et al.*, 1994).

**Experimental sedimentology**: Important laboratory experiments were carried out on antidune structures (Yagishita and Taira, 1989; Yagishita *et al.*, 1992; Yokokawa *et al.*, 1999) and on the structures of combined flow ripples (Yokokawa, 1995).

**Sequence stratigraphy**: Sequence stratigraphy, established in studies of the oil geology of the Gulf of Mexico region in the late 1980s, was rapidly developed in Japan in the 1990s (Ando, 1990a, b; Saito, Yo., 1991, 1994;

Okada, 1992; Hoyanagi, 1992; Ito, 1992, 1999; Murakoshi and Masuda, 1992; Okazaki and Masuda, 1992; Masuda, 1993; Arato, 1993, 1994a, b; Takano, 2000). Recent major achievements in sequence stratigraphy were summarized by Saito, Yo. *et al.* (1995).

**Organic sedimentology and the study of diagenesis**: It was A. Aihara and M. Akiyama who made every effort to promote the systematic study of organic matter in sediments. They regard organic matter as a chemical fossil, emphasizing carbon elements for isotope analysis, biomarkers and kerogens as the source of important information for understanding the thermal structure of sedimentary basins, the history of basin development and the sedimentary environments of basins (Akiyama, 1995a, b). Diagenetic studies of carbonates were carried out by Matsuda (1995, Matsuda *et al.*, 1995). The shallow-sea carbonate sequence and its diagenesis were also studied by Matsuda (1995) and Matsuda *et al.* (1995). Matsumoto (1991) summarised recent achievements of diagenetic studies and developed further the study of gas hydrates (Matsumoto, 1995).

**Sea-floor environments**: Varieties of studies were carried out: Shelf facies analysis and sea-floor environmental analysis off Sendai, northern Honshu (Saito, Yo., 1989, 1994), palaeoenvironments of the Sea of Japan (Oba *et al.*, 1991; Tada, 1994), submarine environments off San-in, southern part of the Sea of Japan (Ikehara, 1991), submarine environments and tectonic developments of the area around Tsushima and Goto Islands, north of Kyushu (Katsura, 1992), dynamic topographic features at the collision zone in the Amami Rise (Kato, 1997) and submarine environments on the slope of the southern Ryukyu Trench (Ujiié *et al.*, 1997).

**Environmental analysis of lacustrine floors**: A systematic study of lacustrine sediments was carried out in Lake Shinji on the Japan Sea side of western Honshu (Inouchi *et al.*, 1990; Tokuoka, 1993, 1999; Tokuoka *et al.*, 1993).

**Tectonic movements and sedimentation**: Research on Cretaceous regional tectonic movements in East Asia and the genesis of sedimentary basins were advanced (Sakai, 1992; Kimura, 1997; Sakai and Okada, 1997; Okada and Mateer, 2000; Okada and Sakai, 2000).

**Sedimentary structures**: A new insight into wave-related sedimentary structures and gravity-flow sedimentary structures was given by Okazaki (1999), Masuda (2001) and Naruse *et al.* (2001). A translation of a book

*Basics of Physical Stratigraphy and Sedimentology* by W. J. Fritz and J. N. Moore (1988), which provided help and support for the observation of sedimentary structures and facies, was completed by K. Harada (1999). Koji Yagishita (2001) published an original textbook entitled *Facies Analyses and Sedimentary Structures* which is very helpful to workers on lithofacies analysis of sedimentary structures and sedimentary environments.

**Theoretical sedimentology**: T. Muto developed theoretical sedimentology, discussing and proposing the retreat theory on the relationship between deltaic progradation and relative sealevel changes (Muto, 1992, 1997; Muto and Steel, 1992). He explains that the delta-front region will undergo an inevitable retreat after a relatively brief period of progradation, resulting from the progressive increase in the effective area of the delta's subaqueous slope, whereby the constant input of sediment does not allow a steady accretion of the slope. Thus, it is not necessary to invoke any change in tectonic and/or palaeoclimatic factors to explain the retreat phases of deltas recognized in the stratigraphic record.

**Trace fossils and sedimentology**: Dynamic studies on the genesis of trace fossils and their relation to environmental changes were developed (Kotake, 1994; Ichihara *et al.*, 1996; Nara, 1997). A Japanese translation of the book *Trace Fossils* by R. G. Bromley (1990) was published by Omori (1993).

### *2000–*

The most important event in this stage was the official recognition of the Sedimentological Society of Japan in the autumn of 2002 as a new academic society under the Science Council of Japan supported by the Japanese Government. This had been the great desire of all sedimentologists since the first organization of the Sedimentology Subcommittee of the Science Council of Japan in 1951. The Society has started with about 600 members.

In the twenty-first century, Japanese sedimentology expects to contribute more to social sedimentology and achieve greater international recognition through links with the International Association of Sedimentologists (IAS) which is scheduled to hold its 17th International Sedimentological Congress (ISC) in Fukuoka, Japan, in 2006. In addition, sedimentologists are also expecting to make a large commitment to the study of deep ocean drilling by the new drilling vessel *Chikyu (Earth)* to be fully commissioned in 2007.

# 9

# Sedimentology in the twenty-first century

In this final chapter three major twenty-first century developments in sedimentology are discussed. It is widely accepted that the twenty-first century is the age of concern for the Earth's environment. The discipline of the geological sciences, particularly sedimentology, will play a very important role in establishing countermeasures and offering answers to the problems facing the global environments many of which are the legacy of science and technology as they developed in the twentieth century. For this purpose we need to ask what sedimentologists can do, and how should sedimentology develop in the future. Three themes will undoubtedly influence the future development of sedimentology: the relationship of sedimentology to stratigraphy, the study of extra-terrestrial sedimentology, and a new approach to environmental sedimentology.

## 9.1. Unification of sedimentology and stratigraphy

As has been repeatedly mentioned, sedimentology, as a new field of the earth sciences, developed in association with stratigraphy in the period from the 1800s to the 1960s as modern geology was established. This circumstance was well illustrated in the classical textbooks on stratigraphy by Grabau (1913), Krumbein and Sloss (1953) and Dunbar and Rodgers (1957). However, based on observations and analyses of strata and sediments which recorded the history of evolution of earth-surface environments and earth systems, a distinct scientific discipline called 'sedimentology' became established as a fundamental science for exploring the history of past environments of the Earth (*see* Figs. 1.1, 1.2). As a consequence, sedimentology has become an important field of the earth sciences for interpreting, among others, palaeoclimatology, palaeogeography, palaeoecology, palaeomarine science and palaeotectonics. The large number of textbooks on sedimentology published after 1960 illustrates both the development and acceptance of this new approach to sedimentology; for example, Potter and Pettijohn (1963), Pettijohn *et al.* (1972), Blatt *et al.* (1972), Reineck and Singh (1973), Pettijohn (1975), Reading (1978, 1986), Miall (1984, 1990) and P.A. Allen and J.R. Allen (1990). In addition, the following are examples of books with 'Sedimentology' in their title: Vatan (1967), Bouma (1962), Berner (1971), Ginsburg (1973), Selley (1976), Kazanskii (1976), Friedman and Sanders (1978), Fairbridge and Bourgeois (1978), Potter *et al.* (1980), Leeder (1982), Friedman and Johnson (1982), Lewis (1984), Allen (1985a, b), Brenchley and Williams (1985), Boggs (1987, 1995), Lindholm (1987), Fritz and Moore (1988), Hsü (1989, 2004), Chamley (1990), Sengupta (1994), Nichols (1999).

This trend, however, began to be further modified from around 1987. This modification has been expressed by the amalgamation of sedimentology and stratigraphy. It was started by Sam Boggs, Jr. (1987), and was followed by Fritz and Moore (1988) and Nichols (1999). Their common concept was that although sedimentology and stratigraphy had developed in their earliest stages as a combined, single field, both became separated by around the 1950s to 1960s. The re-unification of sedimentology and stratigraphy is a natural development,and it seems that this new movement was, in part, stimulated by the establishment of sequence stratigraphy which developed rapidly in the 1970s to 1980s. This trend is the product of the development of seismic stratigraphy which has made it possible to determine the three-dimensional geometry of strata, and to the establishment of precise chronostratigraphy mainly due to developments in biostratigraphy (Guex, 1991) and in radiochronology (Williams *et al.*, 1988; Hole, 1998).

Sedimentology is essentially the discipline for determining the physical, chemical and biological characters of sediments and sedimentary rocks, elucidating the genesis and processes of their formation, and establishing the environments at the time of their deposition. Stratigraphy provides us with the sequential order of sediments and sedimentary rocks and their time correlation. Recent advances in stratigraphy are exhibited by Doyle and Bennett (1998), as in the development of event stratigraphy (Einsele, 1998), cyclostratigraphy (Gale, 1998), chemostratigraphy (McArthur, 1998) and magnetostratigraphy

(Hailwood, 1989). This trend is also being promoted in the new concept of 'experimental stratigraphy' (Paola *et al.*, 2001). Paola and colleagues have attempted to understand quantitatively the mechanics of basin subsidence and sedimentation through an experimental basin, their 'experimental EarthScape (XES)' in which they attempted to test experimentally stratigraphic models reflecting sequence stratigraphy.

Sedimentology and stratigraphy are closely related to each other and should be developed in harmony. In particular, for the purpose of the quantitative analysis of depositional processes in sedimentary basins as well as for the prediction of stratigraphic patterns, a closer relationship between sedimentology and stratigraphy can be expected in the future.

## 9.2 Development of extra-terrestrial sedimentology

A splendid achievement representing the significance of science and technology in the late twentieth century was the start of the space development project launched as one of the activities for the International Geophysical Year of 1957. This has led to successful, if yet preliminary, investigations of Earth's Moon and the Solar planets by NASA (National Aeronautics and Space Administration, USA) through the Pioneer Project (started in 1958; missions by *Pioneer 10* to Jupiter and by *Pioneer 11* to Saturn), the Apollo Project with manned expeditions to the Moon in 1960–1972, the Mariner Project (missions to Mars, Venus and Jupiter started in 1962), the Viking Project (soft-landing missions to Mars in 1975–1976), the Voyager Project (missions to Jupiter, Saturn, Uranus and Neptune in 1977–1989), and the Mars Exploration Rover (MER) mission in 2003/04.

In particular, *Apollo 11* launched by the Apollo Project made the successful landing on the Moon on 20 July 1969, and two astronauts left their footprints on the lunar surface for the first time. It was indeed an inspiring achievement. Since then, studies of sediments on the surface of the Moon and the planets other than the Earth have progressed, as shown in the missions to Mars (the Mariner Project) and Venus (the Viking Project), and sedimentological studies of lunar and martian sedimentations were quickly developed (Mutch *et al.*, 1976; Lindsay, 1976, 1978; Sharp, 1978). Thus, a new field called 'extra-terrestrial sedimentology' was added to sedimentology in the 1970s. At the close of 2003 and the beginning of 2004, NASA's Mars Exploration Rover (MER) mission successfully landed two vehicular robotic instrument packages (*Spirit* and

*Opportunity)* which can be manoeuvred on the Martian surface. Their task is to search for evidence of aqueous environments and signs of past or even present life (Bell, 2003; Sullivan and Bell, 2004; Lane, 2004). Furthermore, renewed explorations of the Moon are planned in the near future by the European Space Agency, Japan and the U.S.A. to solve lingering lunar mysteries (Spudis, 2004). It seems that the field of extraterrestrial sedimentology will become more significant in the future, especially if, as has been suggested, human beings will land on the Moon and Mars in this century.

In this section, recent progress in extra-terrestrial sedimentology is summarized briefly. Whether sedimentary processes resulted from the existence of water and/or some other liquid, or under dry environments is an interesting and controversial topic. As an example of a no liquid scenario, lunar sedimentology is examined, and for the possibility of a liquid involvement, martian sedimentology is discussed.

### 9.2.1 Lunar sedimentology

When the *Apollo 11* mission successfully landed on the surface of the Moon, our eyes were focused on the clear marks of the astronauts' footprints (Fig. 9.1). This strong image suggested to us the development of considerable amounts of unconsolidated sediments on the lunar surface possibly the product of high energy activity. This implies the acceptance of lunar sedimentology regardless of the absence of water (Lindsay, 1972, 1974, 1976).

Some unique physical properties of the lunar surface include a gravity of 1/6th that on Earth's surface and surface temperatures of more than 100° in the daytime and of -150° at night. There is no atmosphere. The magnetic field is, in general terms, negligible, though local magnetic anomalies of unknown origin exist.

#### *Morphology related to sedimentation and erosion*

The most characteristic morphology of the lunar surface is its craters or basins and the 'seas' (*maria*). It is considered that most of the former were produced by high-velocity meteorite impact, and not by volcanic processes as was once believed. According to Spudis (1996), a circular or polygonal depression with diameter of more than 300 km on the lunar surface should be called a basin. Large-scale basins sometimes show overlapping depressions with concentric rings. On and around the soma of these craters and basins are found thick sediments, which attain a width as great as 1000 km. Elevated, highland areas with crater morphology are mostly covered with clastic breccias and melted breccias.

**Fig. 9.1** Footmarks of astronauts clearly printed on the lunar surface (Spudis, 1996). An *Apollo 11* astronaut stood on the Moon on 20 July, 1969. (*Reproduced by permission of* Scientific American,)

Another example of significant lunar morphology is the sinuous rills comparable to erosional river topography on the Earth's surface. They are regarded as channels eroded on basement rocks by the action of turbulent flows of lavas. Generally, sinuous rills start from craters and end in the maria. In addition, breccia gravels similar to dreikanters, which are wind-worn stones commonly found in deserts on the Earth, are observable. These gravels are considered to have been created by the impact of small meteorites (Spudis, 1996).

As mentioned above, a great deal of erosional morphology on the lunar surface seems to be due to shock modification by meteorite impact and related base surges.

### *Regolith*

Ejecta produced by impact pressure at the time of crater genesis on the lunar surface are considered to have covered the lunar surface as a blanket. Sediments produced in this way are called regoliths. They are lunar clastics. The upper sequence, as thick as a few kilometres of the lunar crust was destroyed by meteorite impacts, generating thick sediments called megaregoliths. Each mission of *Apollo 11* (July 1969), *Apollo 12* (November 1969), *Apollo 13* (1970) and *Apollo 14* (1971) collected large amounts of regolith samples.

According to the information of *Apollo 11,* the thickness of regoliths in the Mare Tranquillitatis was about 3 to 6 m, of which the upper 2 to 3 cm was dug up easily by a hammer, but this method failed to dig deeper than 15 cm. Generally, the thickness of regoliths is 2 to 8 m in the lower region called the *maria*, whereas that in the highland region is 20 to 30 m. In addition, the thickness of regoliths tends to increase with time and as the cumulative total of meteorite collisions increases.

Regoliths are divided into lunar soil (Figs. 9.2, 9.3), soil breccias and crystalline breccias (Lindsay, 1978). Lunar soil consists of unconsolidated fine-grained sediments with large and small breccias, covering all the surface. Fine-grained sediments with ill-sorted particles of 3.5ø on average are considered to have been formed by micrometeorite impact with hypervelocities of more than 15 km/sec which destroyed lunar surface rocks and generated fine-grained particles.

Breccias consist of soil breccias and crystalline breccias (Lindsay, 1972, 1974). The former is characterized mainly by dark-coloured, highly porous breccia fragments solidified by a glassy matrix. The latter consists of rounded grains of plagioclase and pyroxene crystals and basaltic rock fragments consolidated with a glassy fine-grained matrix. All these breccias are the products of impact-generated base surge. Impact waves at the time of collision of meteorites on the lunar surface with hypervelocities of more than 10 km/sec show the effects

**Fig. 9.2** Lunar soil (Lindsay, 1978). A glassy grain at the lower left is 3.3 cm in diameter. (*Reproduced with permission of Elsevier.*)

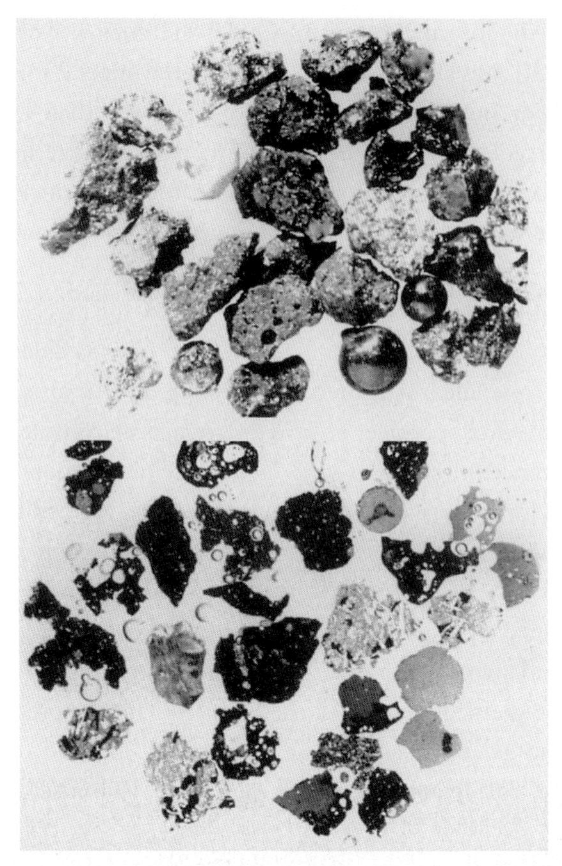

**Fig. 9.3** Photomicrographs of lunar soils composed of rock fragments and glass (Lindsay, 1978). Upper: stereoscope photo; lower: thin-section photo in ordinary light.

of strong compression at the first moment and then quick expansion. At this moment materials exhibit impact evaporation and melt, and most regoliths turn into glasses. In general, the lunar soil is produced by impact melting, and its distribution is limited. The textures of crystalline breccias suggest that they were produced by large-scale base surges due to impact events. Thus, when breccias are exposed repeatedly to many impact base surges, breccias of many generations are produced.

### *Mineral composition*

According to the samples collected by the *Apollo 15* mission, regolith sediments consist of 5–11% plagioclase, 2–16% pyroxene (orthopyroxene + clinopyroxene) together with mineral grains of olivine, spinel and opaque minerals, basalt fragments and glassy matrix (25%). Plagioclase grains are often well rounded and show shock modification. Since kinetic energies on the lunar surface are generated by impact events, the maturity of the lunar-surface sediments is shown by the glass-formation rate and amounts of fine particles, as indicated by an expression (glass + matrix) / (minerals + rock fragments + glass + matrix) (Lindsay, 1972).

### 9.2.2 Martian sedimentology

The spacecraft *Mariner 9* launched in May 1971 has sent back to Earth a great deal of surface-morphology information from Mars (Murray, 1973). The important general feature of Mars is that it is divided into a Northern Hemisphere and a Southern Hemisphere, and that the Southern Hemisphere is characterized by innumerable crater morphologies crossed by large-scale channel-like depressions, whereas the Northern Hemisphere is characterized by a flat topography, and though distinct volcanic craters are present they are limited in number. In particular, the mission of *Viking 2* successfully performed a soft landing on the Martian surface on 3 September 1975 and revealed with clear photographs vast areas covered by gravel deposits (Arvidson *et al.*, 1978) (Fig. 9.4). *Viking 2* has continued to send important clear photographs of the Martian surface, and more than six million pictures had been received by March 1982. Furthermore, *Mars*

Fig. 9.4. Photograph by *Viking Lander 2* at Utopia Planitia on Mars showing the remarkably flat terrain with a field of boulders partly embedded in finer-grained material (Arvidson *et al.*, 1978). Scale is shown by the largest block in the middle right, which is 2.75 m from the spacecraft and is 35 cm wide. (*Reproduced by permission of the Geological Society of America.*)

*Global Surveyer (MGS)* which arrived in orbit on 12 September 1997 has also provided many precise photographs. Based on such valuable information Martian sedimentology is being established. At the beginning of 2004 two unmanned rovers named *Spirit* and *Opportunity* landed on Mars. Their purpose is to survey limited areas of the Martian surface, sample the surface and near surface rocks, carry out *in situ* analyses, search for water and seek evidence of past and, possibly, present life (Lane, 2004; Musser, 2004).

### *Sediments*

The most impressive stratified sediments are found in the polar regions. They extend to an area of roughly 1.6 x $10^6$ $km^2$ around the South Pole, and 1.1 x $10^6$ $km^2$ around the North Pole. The maximum thickness of these layers is about 2~4 km, and the average is about 30 m. They are remarkably uniform in thickness over extensive areas. Murray *et al.* (1973) mainly attribute this uniformity to variations in the obliquity of the spin axis of Mars. The sediments are considered to consist of water ice, solid $CO_2$, $CO_2$-clathrate, dust and volcanic ash. Volcanic activity seems to be widespread and has supplied significant amounts of pyroclastic materials (Carr, 1973).

The Mars Exploration Rovers (MER) *Spirit* and *Opportunity* have already found coarse-grained hematite (>10 microns in diameter) which may have been formed in water on Mars (Lane, 2004). It is expected that the origin and environment of the Martian hematite will soon be clarified.

### *Erosional processes*

Very impressive topographic features on the Martian surface are quite similar to erosional topographies and braided channels on the Earth's surface (Figs. 9.5, 9.6). Large-scale channel structures are considered to have been produced by the action of running water (Milton, 1983; Craddock and Maxwell, 1993). Many cases show that, near the top of large-scale channel walls, a number of more recently formed, smaller channels are developed (Fig. 9.5 right). Morphological features of individual smaller channels indicate that each channel is V-shaped in section, wider and deeper in the upper slope and tapers off downslope. This fact suggests that the Martian surface, completely dried at present, was eroded by liquid water which emerged by seepage. The average temperature of the Martian surface is very low, as cold as -63°. Why, then, is the frozen ice melted and why does it flow? Jakosky (2000) answered this question by suggesting that the frozen ice was melted by geothermal heating some 1~3 km below the surface, and the

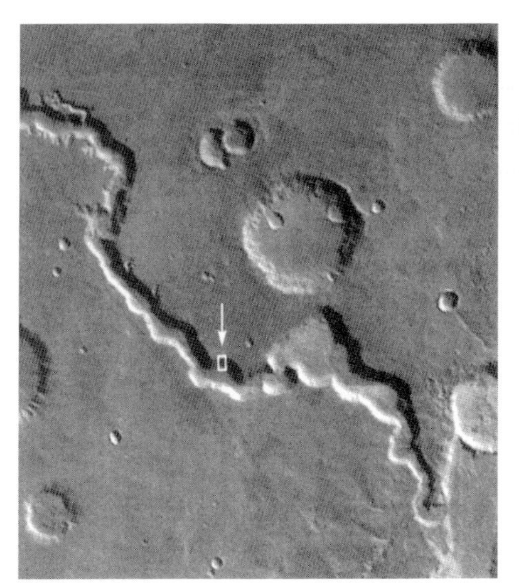

**Fig. 9.5** An ancient valley in Nirgal Vallis on Mars (left) and many channels nearly 1 km long running down the south-facing slope (right) (Jakosky, 2000). The small white box indicates the location of many smaller channels shown in the right-hand image.

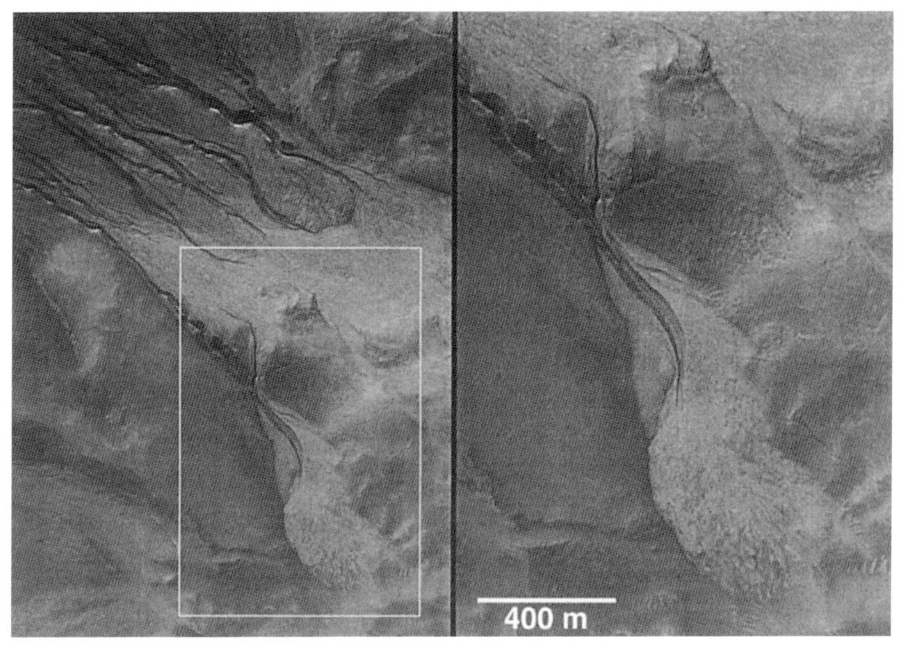


**Fig. 9.6** The high-resolution image by the Mars Orbiter Camera (MOC) showing numerous channels and apron-shaped debris deposits in the Gorgonum Chaos region on Mars (left) (Jakosky, 2000). The debris flow first created the depositional apron and then additional debris and water flows cut the trenches into the apron (right). These features are considered to be extremely recent in their formation (Jakosky, 2000). (*Reproduced by permission of the Planetary Society.*)

water flow rapidly caused erosional structures on the Martian surface but was soon frozen by low temperatures and the erosive processes ceased.

In addition, sub-aerial erosion is more common in the polar regions (Cutts, 1973). According to Hynek and Phillips (2001), extensive denudation of the Martian highlands took place in an interval of 350–500 Ma before present, which is comparable to the denudation of typical slopes in temperate maritime climates on the Earth. There is ample evidence of high velocity dust storms on Mars which must have powers of erosion and can cause the redistribution of debris (Arvidson, 1972).

### *Sedimentary topography*

At the foot of precipitous cliffs of craters and troughs abundant mass-movement deposits are found. Figure 9.6 shows an occurrence of debris-flow deposits, a kind of mass-movement deposit. In particular, the northern plains of Mars are covered by vast deposits of debris flows, which may be as thick as 2–3 km in the centre of the deposit. These deposits cover about 2.7 x $10^7$ $km^2$, or about one-sixth of the total surface of Mars, and the elevation of the deposits ranges up to 4700 m (Tanaka *et al.*, 2001). As an active agent of debris flows, Tanaka *et al.* (2001) interpreted that pore-filling subsurface $CO_2$ or $CO_2$-$H_2O$ mixtures enhanced propagation of debris flows. In this respect, the origin of mass-movements on Mars differs from that on the Earth. Mars has many dune fields of huge size, one of which is 30 by 60 km in area (Cutts and Smith, 1973). They appear to have been produced by the strong aeolian activity.

## 9.3. Environmental sedimentology and social sedimentology

Important and significant as the unification of sedimentology, stratigraphy and extraterrestrial sedimentology may be, there is a more critical issue to be faced by future sedimentologists. It is concerned with no less than the future quality of life of Earth's inhabitants – indeed it concerns the future existence of the human race. It is the environmental crisis of the Earth (Carson, 1962; Colborn *et al.*, 2001). This crisis can be regarded as an inevitable result of human activity due to the rapid increase in human population and the recent rapid development of human industrialized civilization (Park, 2001). Therefore, the concept of earth environment should be better divided into two issues; one is the earth environment as controlled by natural phenomena during the 4600 m.y.-long history of the Earth and the other is that caused by human activity. As hitherto stated, sedimentology is the science of the earth's environment, and it naturally follows that the objectives of sedimentological studies are also divided into the natural products of the earth's environment and the artificially produced environments. It is proposed, therefore, that sedimentology be subdivided, for the sake of convenience, into environmental sedimentology, which deals with the naturally produced earth environments, and social sedimentology, which deals with man-made earth environments.

It is widely accepted that environmental sedimentology is concerned with clarifying natural environmental changes in the earth history recorded in sediments and sedimentary rocks in order to explain past and present earth environments. This is the discipline of sedimentology as described systematically in this book. On the other hand, social sedimentology is to pursue, through sedimentology, environmental changes which are the product of human activities, to clarify problems, to contribute to the conservation of natural environments and to the improvement of environments under threat from human activities. The planning and strategy for the conservation of various environments as proposed by O'Halloran *et al.* (1994) provide good examples of the subject of social sedimentology. At the Malvern International Conference on Geological and Landscape Conservation held in Malvern in the Welsh borderland in 1993, they summarized sustainability, landscape conservation and community initiatives, site conservation and public awareness, and the international significance of the Earth's environmental issues. As Dan Bosence (2001), professor of Royal Holloway College, University of London, described in his report to the 21st IAS meeting held in September 2001 at Davos, Switzerland, 'The most obvious gaps to my mind are in Environmental and Industrial Sedimentology', and he pointed to the importance of these fields. Bosence's 'Industrial Sedimentology' seems to keep in mind mainly hydrocarbon development, but it is clear that as presently defined, 'Industrial Sedimentology' contains both environmental sedimentology and social sedimentology.

As major subjects to which social sedimentology can contribute much, such diverse topics as depositional problems of sediment in dammed lakes, harbours, artificial channels and canals, scouring below dams and increasing risks of coastal erosion due to dam construction, crude oil pollution (Ohshima *et al.*, 1975), diffusion and sedimentation of chemically-polluted materials (Ohshima *et al.*, 1982; Ohshima and Saito, 1993), soil erosion, conservation of lake-bottom environments (Tokuoka *et al.*, 1993; Tokuoka, 1999), and problems of finding geologically stable sites for radioactive nuclear waste repositories (Shimazaki *et al.*, 1995), make only a partial list, but together affect millions of people.

A good example of researches related to social sedimentology was recently shown by K. Oki and his co-workers on the mercury content in surface sediments in an inland sea called the Yatsushiro Sea off the west coast of central Kyushu, Japan (Fig. 9.7) (Rifardi *et al.*, 1998; Tomiyasu *et al.*, 2000). Mercury-contaminated effluent had been discharged into Minamata Bay in the southern part of the Yatsushiro Sea from an acetaldehyde producing factory in Minamata on the west

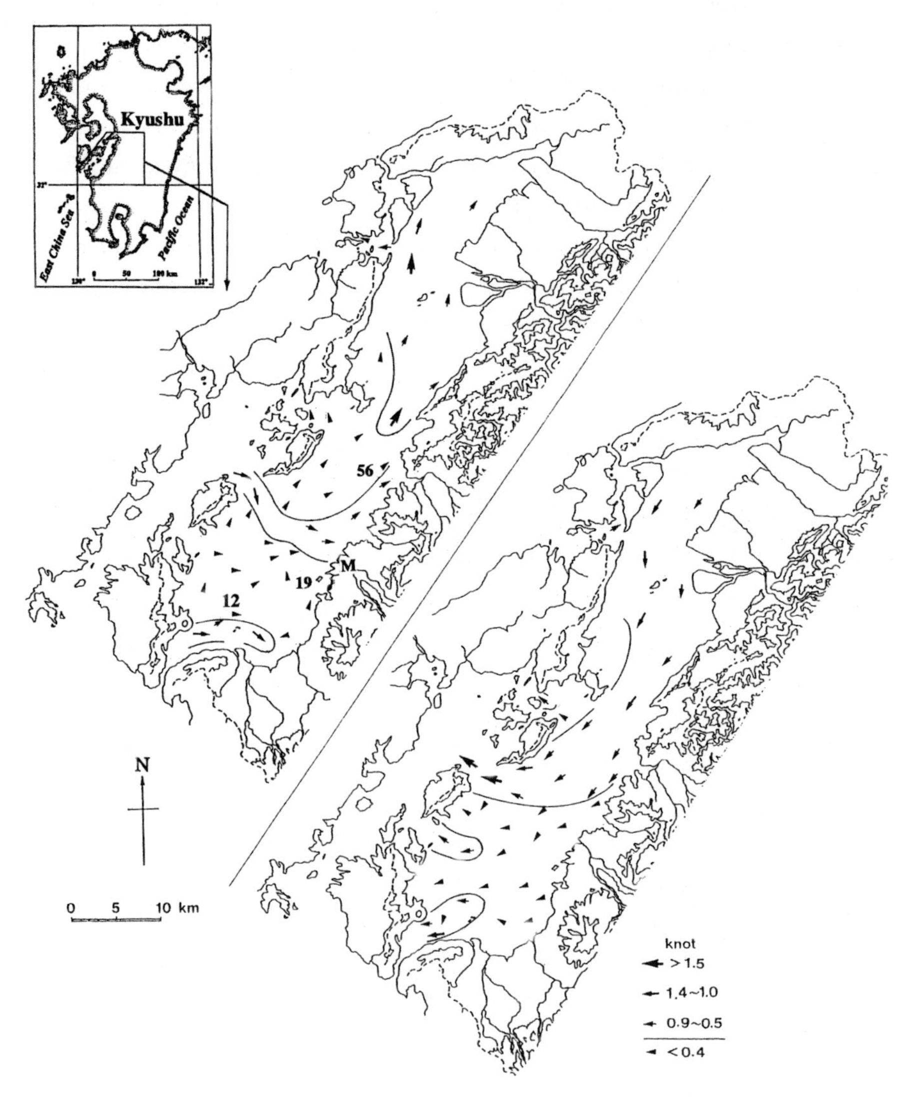


**Fig. 9.7** Index map of the Yatsushiro Sea, central Kyushu in the upper left corner and maps showing velocity and direction of tidal currents (Rifardi *et al.*, 1998). On the upper map M indicates Minamata City and the numbers 12, 19 and 56 in the lower left indicate the sampling stations shown in Fig. 9.8.

coast of central Kyushu from 1956 to 1965. This mercury collected in the bottom mud, was bioaccumulated by fish and shellfish, and caused the outbreak of the serious Minamata disease in humans in 1956 (Harada, 1995; Ninomiya *et al.*, 1995). Most of the polluted sediment in the bay was dredged and removed from 1980 to 1989, but the sediments in the Yatsushiro Sea were not dredged and thus retain a stratified record of the depositional history, as shown below.

Cored samples show that the mercury contents of less than 0.1 ppm (0.036–0.094 ppm) in the deepest part of the cores represent the normal chemical situation of the Yatsushiro Sea (Fig. 9.8). In contrast, the upper parts of the cores show mercury contents of much more than 0.1 ppm (the maximum value is 3.457 ppm at Station 19) (Fig. 9.8). The first appearance of high mercury content at each core-site indicates the arrival of the pollutant at those points. Each sampling point shows a time-lag depending on the distance from the factory and current systems in the sea. The time-lag length is roughly equivalent to the period of about 50 years from 1946 to 1996. Judging from the distribution patterns of the maximum mercury content, the fine-grained sediments polluted by mercury

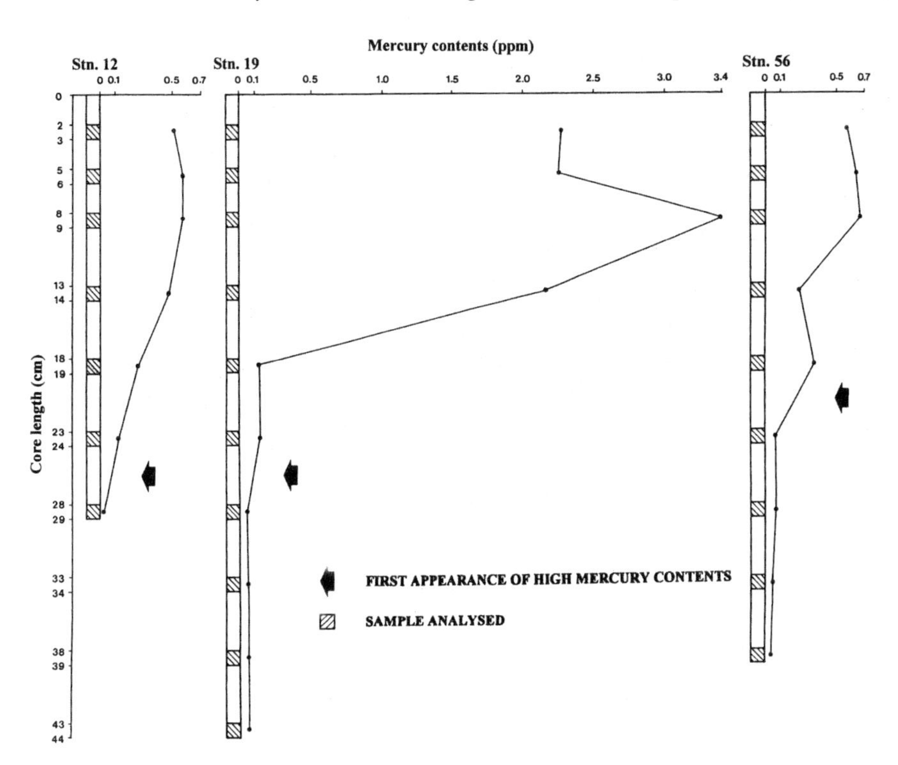


**Fig. 9.8** Vertical distribution of mercury contents in core samples in the southern Yatsushiro Sea (Rifardi *et al.*, 1998). The sampling stations are shown in Figure 9.7.

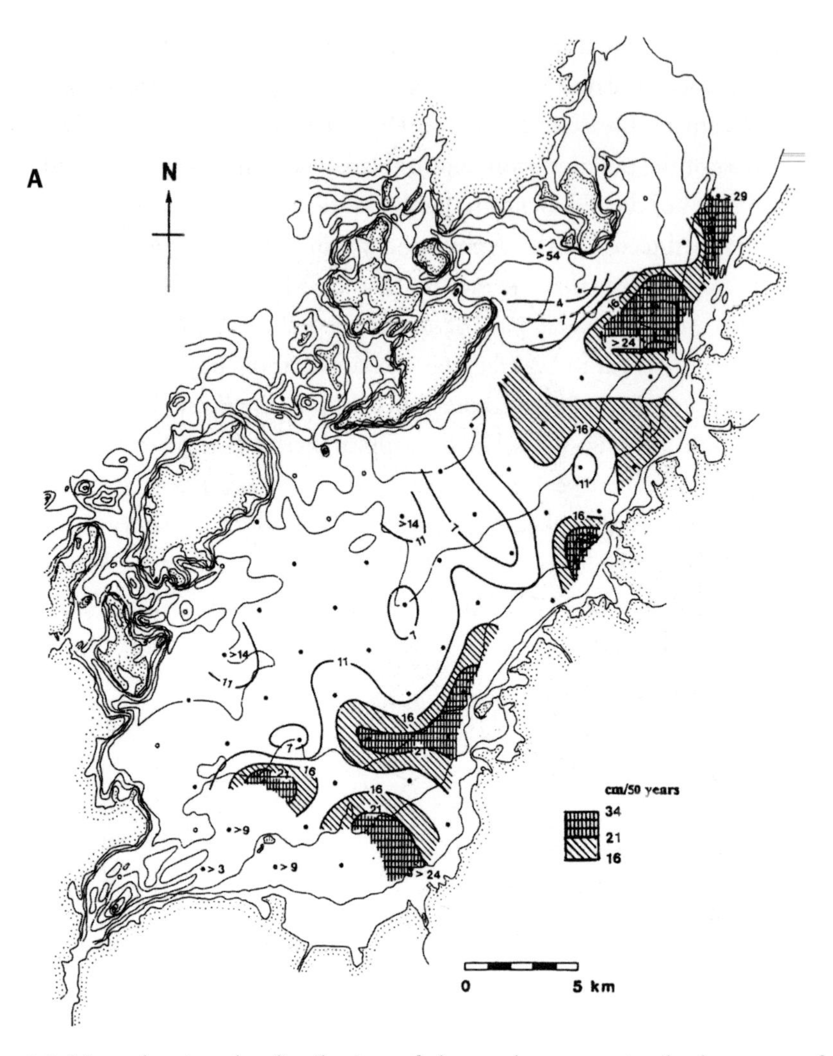


**Fig. 9.9** Maps showing the distribution of the mud contents in the bottom surface sediments (A, *above*) and the distribution of sedimentation rates (B, *opposite*) in the southern Yatsushiro Sea (Rifardi *et al.*, 1998).

were transported both northeastward and southward by longshore currents and spread north and west across the sea in the northern and southern parts of the southern Yatsushiro Sea in harmony with the current system (Figs. 9.9, 9.10). Such sedimentological information provides important documentation about chemical risks for human health. Now, the environmental impact of discharged mercury has become a matter of world-wide concern (Nriagu *et al.*, 1992; Ikingura and Akagi, 1996). Clearly, the case study of the mercury-polluted sediment that caused the Minamata disease is important for elucidating the behaviour

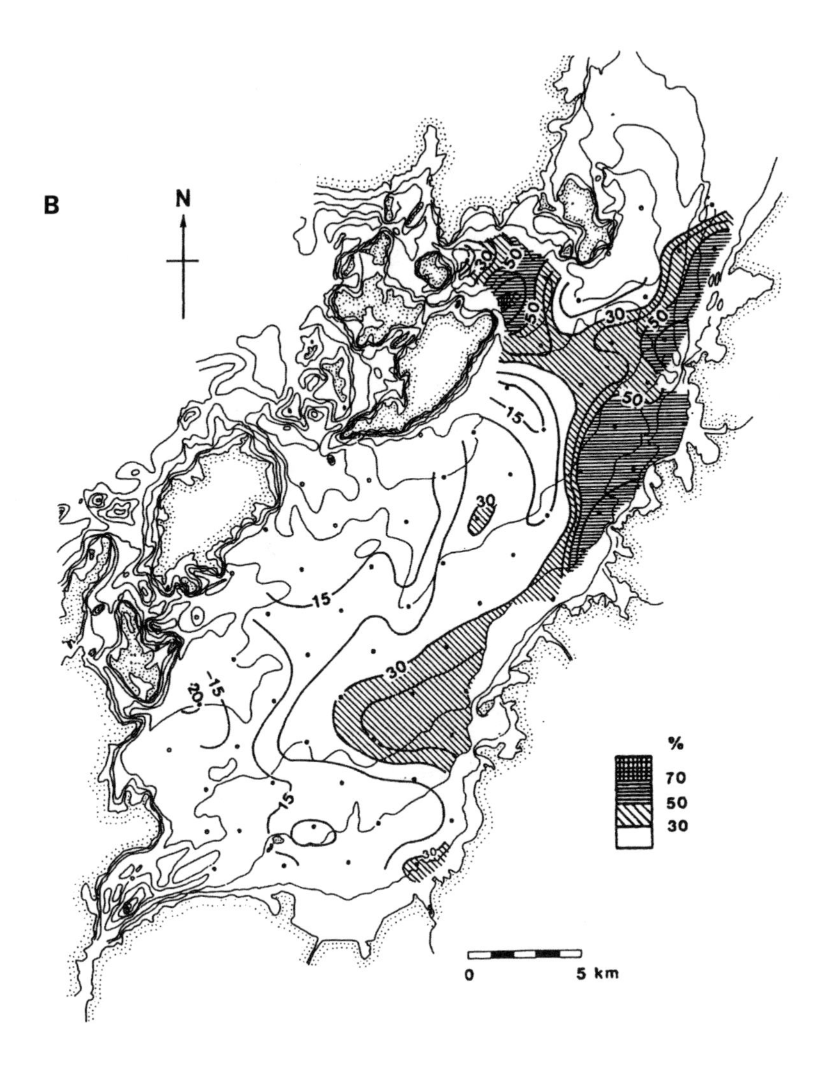


of mercury over a long period. It also gives clues to how sedimentologists may predict the movement of pollution in future events and thus reduce the impact on human and animal life.

The International Association of Sedimentologists (IAS) has only begun to include social sedimentology into its activities fifty years after its establishment (Tucker, 2002). This will be a natural evolution in the history of sedimentology which frequently has been developed in a close relation with the needs of society. This welcome event was first initiated by the 'IAS/SEPM Environmental Sedimentology Workshop on Continental Shelves' held in January 2002 in Hong Kong, and will also be raised as one of the major technical sessions at the 17th International Sedimentological Congress to be held in Japan in 2006.

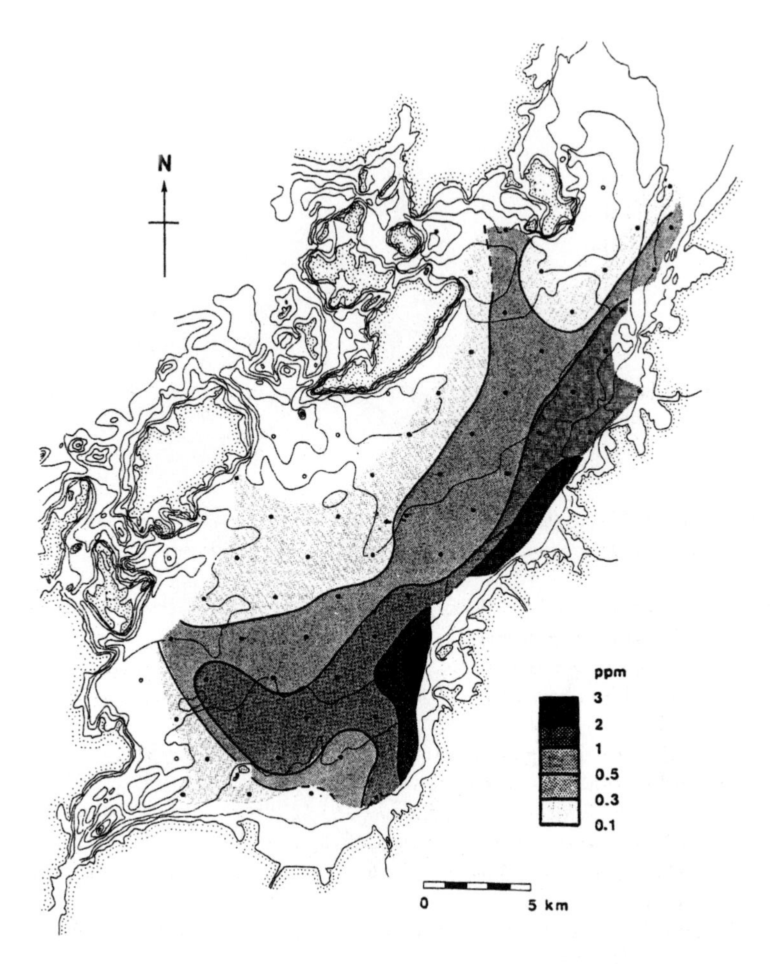


**Fig. 9.10** Map showing the distribution of the maximum mercury content (ppm) in the southern Yatsushiro Sea (Rifardi *et al.*, 1998).

Sedimentology in the twenty-first century must continue to play an important role as a basic discipline in geological sciences, but also will have to contribute more to the conservation and the improvement of Earth's environments. Sedimentologists and geologists have studied the history and environmental changes of the Earth and therefore are well placed to contribute their knowledge to establish the best global environment for future generations. In order to do this effectively, sedimentologists should interact with large and small communities as much as possible as well as developing an international and interdisciplinary perspective. The past history of sedimentology, as presented in this book, suggests that sedimentologists are capable of playing this role.

# References

Publications in Japanese are denoted by (J). Publications in Japanese with an English abstract are denoted by (J+E).


Adachi, M. (1985), The chemical composition of garnets from gneiss clasts in the Kamihirose conglomerate and the Kamiaso conglomerate. *Bull. Mizunami Fossil Museum*, (12), 183-196. (J+E)

Adams, J. A. S. and Weaver, C. E. (1958), Thorium-to-Uranium ratios as indicators of sedimentary processes: Examples of geochemical facies. *Bull. Am. Assoc. Petrol. Geol.*, **42**, 387-430.

Ager, D. V. (1973), *The Nature of the Stratigraphic Record.* Macmillan, London.

_____ (1993), *The New Catastrophism. The Importance of the Rare Event in Geological History.* Cambridge Univ. Press, Cambridge.

Aihara, A. (1983), Coal: geological significance of the characteristic nature of the Japanese Tertiary coal – a case study of applying coal petrology to earth science. *UNESCO Regional Workshop on Coal Geology in SE Asia Final Report, Seoul Natnl. Univ.*, 71–105.

Aitken, J. F. (1994), Coal in a sequence stratigraphic framework. *Geoscientist*, **4**, (5), 9-12.

Akiyama, M. (1995a), *Revival of Molecular Fossils – Introduction to Organic Geology.* Kyoritsu Pub. Co., Tokyo. (J)

_____ (1995b), Organic geology: An invitation to the study of organic matter in geology. *J. Geol. Soc. Japan*, **101** (12), 990–998. (J+E)

Allen, J. R. L. (1964), Studies in fluviatile sedimentation: six cyclothems from the Lower Old Red Sandstone, Anglo-Welsh basin. *Sedimentology*, **3**, 163-198.

_____ (1965a), A review of the origin and characteristics of recent alluvial sediments. *Sedimentology*, **5**, 85-91.

_____ (1965b), Late Quaternary Niger delta and adjacent areas; sedimentary environments and lithofacies. *Bull. Am. Assoc. Petrol. Geol.*, **49**, 547-600.

_____ (1965c), Coastal geomorphology of eastern Nigeria: Beach-ridge barrier islands and vegetated tidal flats. *Geol. Mijnbouw*, **44** (1), 1-21.

_____ (1971), Mixing at turbidity current heads and its geological applications. *J. Sed. Petrol.*, **41**, 97-113.

_____ (1985a), *Principles of Physical Sedimentology.* Allen & Unwin, London.

_____ (1985b), *Experiments in Physical Sedimentology.* Allen & Unwin, London.

_____ (1993), Sedimentary structures: Sorby and the last decade. *J. Geol. Soc. London*, **150**, 417-425.

Allen, P. (1959), The Wealden environment: Anglo-Paris basin. *Phil. Trans. Royal Soc. London*, **242 B**, 283-346.

Allen, P. A. and Allen, J. R. (1990), *Basin Analysis: Principles and Application.* Blackwell, Oxford.

Anderton, R., (1985), Clastic facies models and facies analysis. *In*: P. J. Brenchley and B. P. J.

Williams (eds.): *Sedimentology. Recent Developments and Applied Aspects*. Blackwell Sci. Publ., Oxford, 31-47.

Ando, H. (1990a), Stratigraphy and shallow marine sedimentary facies of the Mikasa Formation, Middle Yezo Group (Upper Cretaceous). *J. Geol. Soc. Japan*, **96** (4), 279-295. (J+E)

_____ (1990b), Shallow-marine sedimentary facies distribution and progradational sequences of the Mikasa Formation, Middle Yezo Group (Upper Cretaceous). *J. Geol. Soc. Japan*, **96** (6), 453-469. (J+E)

Aoyagi, K., Sato, T. and Kazama, T. (1970), Distribution and origin of the Neogene rocks in the Akita oil fields, Japan. *J. Japan. Assoc. Petroleum Geologists*, **35** (2), 67-76. (J+E)

Aoyama, N. (1958), *Petrography [IV]*. Kobutsu Shumino Kai, Kyoto. (J)

Arai, J. (1960), *The Tertiary System of the Chichibu Basin, Saitama Prefecture, central Japan. Part I. Sedimentology*. Japan Soc. Prom. Sci., Tokyo.

Arai, S. and Okada, H. (1991), Petrology of serpentine sandstone as a key to tectonic development of serpentine belts. *Tectonophysics*, **195**, 65-81.

Arato, H. (1993), Sequence stratigraphy and petroleum exploration. Part 1: Overview. *Development and Stockpiling of Petroleum*, **26** (6), 97-114. (J)

_____ (1994a), Sequence stratigraphy and petroleum exploration. Part 2: Analytical techniques. *Development and Stockpiling of Petroleum*. **27** (1), 86-104. (J)

_____ (1994b), Sequence stratigraphy and petroleum exploration. Part 3: Applications and case studies. *Development and Stockpiling of Petroleum*. **27** (2), 58-82. (J)

Arbentz, P. (1919), Probleme der Sedimentation und ihre Beziehungen zur Gebirgsbildung in den Alpen. *Naturforsch. Gesell. Zurich, Vierteljahresschrift, Jahrg.*, **64** (1/2), 246-275.

Argand, E. (1916), Sur l'arc des Alpes Occidentales. *Eclogae Geol. Helv.*, **14**, 145-191.

Armentrout, J.M. (1990), *Integrated Stratigraphic Analysis. Work Book 2. Seismic Facies*. SEPM Short Course, Dallas, Texas, 31-38.

Arrhenius, C. (1963), Pelagic sediments. *In*: Hill, M. N. (ed.), *The Sea. Vol. 3*. Wiley-Interscience, New York, 655-727.

Arthur, M. A., Jenkyns, H. C., Brumsack, H. J. and Schlanger, S. O. (1990), Stratigraphy, geochemistry and paleoceanography of organic carbon-rich Cretaceous sequence. *In*: Ginsburg, R. N. and Beaudoin, B. (eds.), *Cretaceous Resources, Events and Rhythms*. Dordrecht, Kluwer Academic Publishers, 75-119.

Arvidson, R. E. (1972), Aeolian processes on Mars: erosive velocities, settling velocities, and yellow clouds. *Geol. Soc. Am. Bull.*, **83**, 1503-1508.

_____ Binder, A. B. and Jones, K. L. (1978), The surface of Mars. *Scientific American*, **238** (3), 76-89.

Asano, K. (1955), Studies on the ecology and sedimentaton of Matsukawa-ura, Soma City, Fukushima Prefecture. *Tohoku Univ., Inst. Geol. Paleont., Contr.*, (45), 1-96; (46), 1-43. (J+E)

Association for Geological Collaboration (1970), *Dictionary of Geological Sciences*. Heibonsha, Tokyo. (J)

Aubouin, J. (1965), *Geosynclines*. Elsevier, Amsterdam.

Bagnold, R. A. (1956), The flow of cohesionless grains in fluids. *Roy. Soc. London Phil. Trans.*, Ser. B, **249**, 235-297.

Bailey, E. B. (1910), Recumbent folds in the schists of the Scottish Highlands. *Quart. J. Geol. Soc. London*. **46**, 586-618.

_____ (1930), New light on sedimentation and tectonics. *Geol. Mag.*, **67**, 77-92.

_____ (1934), West Highland tectonics: Loch Leven to Glen Roy. *Quart. J. Geol. Soc. London*. **90**, 462-525.

_____ (1936), Sedimentation in relation to tectonics. *Geol. Soc. Am. Bull.*, **47**, 1713-1726.

Barker, E. T. (1976), Distribution, composition, and transport of suspended particulate matter in the vicinity of Willapa submarine canyon, Washington. *Geol. Soc. Am. Bull.*, **87**, 625-632.

Barrell, J. (1906), Relative geological importance of continental, littoral, and marine sedimentation. *J. Geol.*, **14**, 316-356, 430-457, 524-568.
_____ (1917), Rhythms and the measurement of geologic time. *Bull. Geol. Soc. Am.*, **28**, 745-904.
Basan, P. B. (ed.) (1978), *Trace Fossils Concepts*. Soc. Econ. Paleont. Mineral., Short Course, (5), 1-201.
Bates, R. L. and Jackson, J. A. (eds.) (1987), *Glossary of Geology. 3rd Ed.* Am. Geol. Inst., Alexandria, Virginia.
Beard, D. C. and Weyl, P. K. (1973), Influence of texture on porosity and permeability of unconsolidated sand. *Am. Assoc. Petrol. Geol. Bull.,* **57**, 349-369.
Beer, R. M. and Gorsline, D. S. (1971), Distribution, composition and transport of suspended sediment in Redondo Submarine Canyon and vicinity (California). *Mar. Geol.*, **10**, 153-175.
Bell, H. S. (1942), Density currents as agents for transporting sediments. *J. Geol.*, **50**, 512-547.
Bell, J. (2003), The human side of Mars rover exploration. *Planetary Report*, **23** (6), 12-17.
Bennetts, K. R. and Pilkey, O. H. (1976), Characteristics of three turbidites, Hispaniola–Caicos Basin. *Geol. Soc. Am. Bull.*, **87**, 1291-1300.
Berger, W. H. and Winterer, E. L. (1974), Plate stratigraphy and the fluctuating carbonate line. *Int. Assoc. Sediment. Spec. Publ.*, **1**, 11-48.
Berner, R. A. (1971), *Principles of Chemical Sedimentology*. McGraw Hill, New York.
Bertrand, M. (1897), Structure des Alpes françaises et récurrence de certain facies sédimentaires. *VI Int. Geol. Congr., Zurich, Compt Rend.*, 163-177.
Betzer, P. R., Richardson, R. L. and Zimmerman, H. B. (1974), Bottom currents, nepheloid layers and sedimentary features under the Gulf Stream near Cape Hatteras. *Mar. Geol.*, **16**, 21-29.
Bhatia, M. R. (1983), Plate tectonics and geochemical composition of sandstone. *J. Geol.*, **91**, 611-627.
Blatt, H., Middleton, G. and Murray, R. (1972), *Origin of Sedimentary Rocks.* Prentice-Hall, New Jersey.
_____ _____ _____ (1980), *Origin of Sedimentary Rocks. 2nd Ed.* Prentice-Hall, New Jersey.
Bluck, B. J. (1965), The sedimentary history of some Triassic conglomerates in the Vale of Glamorgan, South Wales. *Sedimentology*, 4, 225-245.
Boggs, S., Jr. (1967), A numerical method for sandstone classification. *J. Sed. Petrol.*, **37**, 548-555.
_____ (1987), *Principles of Sedimentology and Stratigraphy*. Merrill, Columbus, Ohio.
_____ (1992), *Petrology of Sedimentary Rocks*. Macmillan, New York.
_____ (1995), *Principles of Sedimentology and Stratigraphy. 2nd Ed.* Prentice-Hall, New Jersey.
Bonney, T. G., 1900: The Bunter pebble beds of the Midlands and the source of their materials. *Quart. J. Geol. Soc. London.* **56**, 286-306.
Bosence, D. (2001), Observer's report on the 21st IAS meeting of sedimentology, Davos, Switzerland 3–5 September, 2001. *IAS Newsletter,* (176), 4-6.
Boswell, P. G. H. (1933), *On the Mineralogy of Sedimentary Rocks*. Murby, London.
Bouma, A. H. (1962), *Sedimentology of Some Flysch Deposits: A Graphic Approach to Facies Interpretation*. Elsevier, Amsterdam.
_____ (1972a), Recent and ancient turbidites and contourites. *Gulf Coast Assoc. Geol. Soc. Trans.,* **22**, 205-221.
_____ (1972b), Fossil contourites in lower Niesen flysch, Switzerland. *J. Sed. Petrol.*, **42** (4), 917-921.
_____ (1978), Bouma sequence. *In*: Fairbridge, R. W. and Boureois, J. (eds.), *The Encyclopedia of Sedimentology*. Dowden, Hutchingson & Ross, Stroudsburg, Pa., 78-81.
Bradley, W. C. (1970), Effect of weathering on abrasion of granitic rocks, Colorado River, Texas. *Geol. Soc. Am. Bull.*, **81**, 61-80.

Bramlette, M. N. (1934), Heavy mineral studies on correlation of sands at Kettleman Hill, California. *Am. Assoc. Petrol. Geol. Bull.*, **18**, 1559-1576.
_____ (1961), Pelagic sediments. *In*: Sears, M. (ed.), *Oceanography.* Publ. Am. Assoc. Advancement Sci., **67**, 345-366.
Brenchley, P. J. and Williams, B. P. (eds.)(1985), *Sedimentology: Recent Developments and Applied Aspects.* The Geological Society, Blackwell, Oxford.
Bretz, J. H. (1963), Memorial to Hakon Wadell. *Proc. Geol. Soc. Am.* (1963), 53-55.
Bridge, J. S. (1993), The interaction between channel geometry, water flow, sediment transport and deposition in braided rivers. *In*: Best, J. L. and Bristow, C. S. (eds.), *Braided Rivers.* Geological Society Special Publication, **75**, 13-72.
Brinkmann, R. (1993), Über Kreuzchichtung in deutschen Buntsandsteinbecken. *Göttingen Nachr., Math. Physik, Kl. IV,* (32), 1-12.
Brock, E. J. (1974), Coarse sediment morphometry: A comparative study. *J. Sed. Petrol.*, **44**, 663-672.
Bromley, R. G. (1990), *Trace Fossils: Biology and Taphonomy. Special Topics in Palaeontology 3.* Unwin Hyman Ltd., London. (J+E)
Brongniart, A. (1821), Sur les caracteres zoologiques des formations, avec l'application de ces caracteres a la determination de quellues terrains de Craie. *J. de Mines*, **6**, 537-571.
Budd, D. A., Saller, A. H. and Harris, P. M. (eds.) (1995), *Unconformities and Porosity in Carbonate Strata.* Am. Assoc. Petrol. Geol. Bull., Mem., (63).
Bull, P. A. (1986), Procedures in environmental reconstruction by SEM analysis. *In*: Sieveking, G. de C. and Hart, M. B. (eds.), *The Scientific Study of Flint and Chert.* Cambridge Univ. Press, Cambridge, 221-226.
Burger, H. and Skala, W. (1976), Comparison of sieve and thin-section techniques by a Monte-Carlo model. *Computer Geoscience*, **2**, 123-139.
Cadet, J. P., Kobayashi, K., Lallemant, S., Jolivet, L., Aubouin, J., Boulègue, J., Dubois, J., Hotta, H., Ishii, T., Konishi, K., Niitsuma, N. and Shimamura, H. (1987), Deep scientific dives in the Japan and Kuril Trenches. *Earth Planet. Sci. Letts.*, **83**, Special Issue, 313-328.
Carr, A. P. (1971), Experiments on longshore transport and sorting of pebbles, Chesil Beach, England. *J. Sed. Petrol.*, **41**, 1084-1104.
Carr, M. H. (1973), Volcanism on Mars. *J. Geophys. Res.*, **78**, 4049-4062.
Carson, R. (1962), *Silent Spring.* Houghton Mifflin Co., Boston.
Cayeux, Lucien (1929a), *Les roches sédimentaires de France. Roches silicieuse.* Imprimerie nationale, Paris.
Cayeux, Lucien (1929b), *Les roches sédimentaires de France. Roches carbonatées.* Mason, Paris.
Chamley, H. (1990), *Sedimentology.* Springer-Verlag, Berlin.
Chang, K. H. (1973), Toward an international guide to stratigraphic classification, terminology, and usage. *J. Geol. Soc. Korea,* **9** (2), 123-125.
_____ (1976), *A Study of Stratigraphic Classification in Korea with Special Reference to Synthems.* Ph.D. Thesis, Princeton University.
Chappell, B.W. (1968), Volcanic greywackes from the Upper Devonian Baldwin Formation, Tamworth-Barraba district, New South Wales. *J. Geol. Soc. Australia,* **15**, 87-102.
Chijiwa, K. (1985), Recognition and sedimentary petrographical features of storm deposits: an example of the Miocene Kumano Group. *J. Sed. Soc. Japan,* (22-23), 100-107. (J+E)
Cipriani, C. (1992), Nicolaus Steno: Geologic fundator. *Acta Vulcanolgica, Marinelli Volume,* **2**, 125-131.
Cloos, H. (1938), Primäre Richtungen in Sedimenten der rheinischen Geosynkline. *Geol. Rundschau,* **29**, 357-367.
Colborn, T., Dumanoski, D. and Myers, J. P. (2001), *Our Stolen Future.* Spieler Agency, Inc., New York.

Correns, C. W. (1939), Die Sedimentgesteine. *In*: *Die Entstehung der Gesteine*. Springer, Berlin, 116-262.

_____ (1949), *Einführung in die Mineralogie*. Springer.

Craddock, R. A. and Maxwell, T. A. (1993), Geomorphic evolution of the Martian highlands through ancient fluvial processes. *J. Geophys. Res.*, **98**, 3453-3468.

Crook, K. A. W. (1960), Classification of arenites. *Am. J. Sci.*, **258**, 419-428.

Cummins, W. A. (1962), The greywacke problem. *Liverpool Manchester Geol. J.*, **3**, 51-72.

Curray, J. R. and Moore, D. G. (1971), Growth of the Bengal deep-sea fan and denudation in the Himalayas. *Geol. Soc. Am. Bull.*, **82**, 563-572.

Cutts, J. A. (1973), Wind erosion in the Martian polar regions. *J. Geophys. Res.*, **78**, 4211-4221.

_____ and Smith, R.S.U. (1973), Eolian deposits and dunes on Mars. *J. Geophys. Res.*, **78**, 4139-4154.

Cuvier, G. (1826), *Discours sur les Revolutions de la Surface du Globe, et sur les Changemens qu'elles ont Produits dans le Règne Animal*. Dufour et d'Ocagne, Paris.

_____ and Brongniart, A. (1811), *Essai sur la Géographie Mineralogique des Environs de Paris, avec Une Carte Géognostique, et des Coupes de Terrain*. Baudouin, Paris.

_____ and Brongniart, A. (1822), *Description Géologique des Environs de Paris*. Dufour et d'Ocagne, Paris.

Cuvillier, J. and Sacals, V. (1951), *Corrélations Stratigraphiques par Microfacies en Aquitaine Occidentale*. E.J. Brill, Leiden.

Daly, R. A. (1936), Origin of submarine "canyons". *Am. J. Sci.*, **31**, 401-420.

Damuth, J. E. and Embley, R. W. (1979), Upslope flow of turbidity currents on the northwest flank of the Ceara Rise: western Equatorial Atlantic. *Sedimentology*, **26**, 825-834.

Dana, J. D. (1873), On some results of the Earth's contraction from cooling, including a discussion of the origin of mountains, and the nature of the Earth's interior. *Am. J. Sci.,* 3rd Ser., **5**, 423-443, 474-475; **6**, 6-14, 104-115, 161-172, 304, 381-382.

Daubrée, A. (1879), *Études synthétiques de géologies expérimentale*. Dunod, Paris.

Davies, T. A., Weser, O. E., Luendike, B. P. and Kidd, R. B. (1975), Unconformities in the sediments of the Indian Ocean. *Nature*, **253**, 15-19.

De Raaf, J. F. M., Reading, H. G. and Walker, R. G. (1965), Cyclic sedimentation in the Lower Westphalian of north Devon, England. *Sedimentology*, **4**, 1-52.

Dewey, J. F. and Bird, J. M. (1970), Mountain belts and the new global tectonics. *J. Geophysical Res.*, **75** (14), 2625-2647.

Dickinson, W. R. (1970), Interpreting detrital modes of graywacke and arkose. *J. Sediment. Petrol.*, **40**, 695-707.

_____ (1971), Plate tectonic models of geosynclines. *Earth and Planetary Sci. Letters*, **10**, 165-174.

_____ (1985), Interpreting provenance relations from detrital modes of sandstones. *In*: Zuffa, G. G. (ed.), *Provenance of Arenites*. D. Reidel Publ., Dordrecht, 333-361.

_____ and Suczek, C. A. (1979), Plate tectonics and sandstone compositions. *Am. Assoc. Petrol. Geol. Bull.*, **63**, 2164-2182.

_____ Beard, L. S., Brakenridge, G. R., Erjavec, J. L., Ferguson, R. C., Inman, K. F., Knepp, R. A., Linberg, F. A. and Ryberg, P.T. (1983), Provenance of North American Phanerozoic sandstones in relation to tectonic setting. *Geol. Soc. Am. Bull.*, **94**, 222-235.

_____ and Rich, E. I. (1972), Petrologic intervals and petrofacies in the Great Valley Sequence, Sacramento Valley, California. *Geol. Soc. Am. Bull.*, **83**, 3007-3024.

Dietz, R. S. (1962), The sea's deep scattering layers. *Scientific American*, **207** (2), 44-67.

_____ (1963), Collapsing continental rises: an actualistic concept of geosynclines and mountain building. *J. Geol.*, **71**, 314-333.

_____ (1965), Collapsing continental rises: an actualistic concept of geosynclines and mountain building: a reply. *J. Geol.*, **73**, 901-906.

_____ and Holden, J. C. (1966), Miogeoclines (miogeosynclines) in space and time. *J. Geol.*, **74**, 566-583.

Doeglas, D. J. 1951a: From sedimentary petrology to sedimentology? *Proc. 3rd. Intern. Congr. Sedimentology*, 15-22.

_____ 1951b: Sedimentology and petroleum geology. *Proc. 3rd World Petrol. Congr., Sec. 1*, 439-445.

_____ 1955: Progress of regional sedimentology. *Proc. 4th World Petrol. Congr., Sec. 1*, 417-426.

_____ 1976: The first fifteen years of International Association of Sedimentologists. *Sedimentology*, **23**, 5-16.

d'Orbigny, A. D. (1842), *Paléontologie française, Terrains oolitiques ou jurassiques*. **1**. Paris.

Dott, R. H., Jr. (1964), Wacke, graywacke and matrix – What approach to immature sandstone classification? *J. Sed. Petrol.*, **34**, 625-632.

_____ (1974), The geosynclinal concept. *In*: Dott, R. H., Jr. and Shaver, R. H. (eds,), *Modern and Ancient Geosynclinal Sedimentation. Soc. Econ. Paleont. Mineral. Spec. Publ.*, **19**, 1-13.

_____ (1978), Tectonics and sedimentation a century later. *Earth-Sci. Rev.*, **13**, 1-34.

_____ (1979), The geosyncline – First major geological concept "Made in America". *In*: Schneer, C. J. (ed.), *Two Hundred Years of Geology in America*. Univ. New Hampshire Press, 239-264.

_____ (1987), Perspectives: something old, something new, something borrowed, something blue – A hindsight and foresight of sedimentary geology. *J. Sed. Petrol.*, **58** (2), 358-364.

_____ (1997), James Dwight Dana's old tectonics – Global contraction under divine direction. *Am. J. Sci.*, **297**, 283-311.

_____ (2001), Wisconsin roots of the modern revolution in structural geology. *J. Geol. Soc. Am. Bull.*, **113** (8), 996-1009.

_____ (2003a), Sedimentary structures as way-up indicators. *In*: Middleton, G. V. (ed.), *Encyclopedia of Sediments and Sedimentary Rocks*. Kluwer Academic Pub., Dordrecht, 627-628.

_____ (2003b), Joseph Barrell (1869-1919). *In*: Middleton, G. V. (ed.), *Encyclopedia of Sediments and Sedimentary Rocks*. Kluwer Academic Pub., Dordrecht, 638-639.

_____ and Shaver, R. H. (eds.) (1974), Modern and ancient geosynclinal sedimentation. *Soc. Econ. Palent., Mineral. Spec. Publ.*, **19**.

Double, I. S., 1924: The petrography of Late Tertiary deposits of the east of England. *Proc. Geol. Assoc.London*, **35**, 332-358.

Doyle, P. and Bennett, M. R. (eds.) (1998), *Unlocking the Stratigraphical Record. Advances in Modern Stratigraphy.* John Wiley & Sons, Chichester.

Drake, C. L., Ewing, M. and Sutton, G. H. (1959), Continental margin and geosynclines: the east coast of North America, north of Cape Hatteras. *In*: Ahrens, L. H., Press, F., Runcorn, S. K. and Urey, H. C. (eds.), *Physics and Chemistry of the Earth*. **3**. Pergamon, Oxford, 110-198.

Drake, D. E. and Gorsline, D. S. (1973), Distribution and transport of suspended particulate matter in Hueneme, Redondo, Newport and La Jolla submarine canyons, California. *Geol. Soc. Am. Bull.*, **84**, 3949-3968.

Dunbar, C. O. and Rodgers, J. (1957), *Principles of Stratigraphy*. John Wiley, New York.

Dzulynski, S. (2001), *Atlas of Sedimentary Structures from the Polish Flysch Carpathyans*. Inst. Geol. Sci., Jagiellonian Univ., Krakow.

_____ Ksiazkiewicz, M. and Kuenen, Ph.H. (1959), Turbidites in flysch of the Polish Carpathians. *Bull. Geol. Soc. Am.*, **70**, 1089-1118.

_____ and Smith, A. J.(1964), Flysch facies. *Ann. Soc. Géol. Pologne*, **34**, 245-266.

_____ and Walton, E. K. (1963), Experimental production of sole markings. *Trans. Edinburgh Geol. Soc.*, **19**, 279-305.

_____ _____ (1964), *Sedimentary Features of Flysch and Greywackes*. Elsevier, Amsterdam.

Eckel, E. (1982), *The Geological Society of America. Life History of a Learned Society*. Geol. Soc. Am. Mem., **155**.

Edgar, N. T., Holcomb, T., Ewing, J. L. and Johnson, W. (1973), Sedimentary hiatuses in the Venezuelan basin. *Init. Rep. DSDP*, **15**, 1051-1062.

Edwards, A. R. (1973), Southwest Pacific regional unconformities encountered during Leg 21. *Init. Rep. DSDP*, **21**, 641-692.

Edwards, D. A. (1993), *Turbidity currents: dynamics, deposits and reversals.* Springer-Verlag, Berlin.

Einsele, G. (1998), Event stratigraphy: Recognition and interpretation of sedimentary event horizons. *In*: Doyle, P. and Bennett, M. R. (eds.), *Unlocking the Stratigraphical Record. Advances in Modern Stratigraphy.* John Wiley & Sons, Chichester, 145-193.

Eittreim, S., Ewing, M. and Thorndike, E. M. (1969), Suspended matter along the continental margin of the North American Basin. *Deep-Sea Res.*, **16**, 613-624.

Ekdale, A. A., Bromley, R. G. and Pemberton, S. G. (1984), *Ichnology: Trace Fossils in Sedimentology and Stratigraphy.* Soc. Econ. Paleont. Mineral., Short Course, (15).

Erbacher, J., Thurow, J. and Littke, R. (1996), Evolution patterns of radiolaria and organic matter variations: A new approach to identify sea-level changes in mid-Cretaceous pelagic environments. *Geology*, **24**, 499-502.

Eskola, P. (1922), The mineral facies of rocks. *Norsk Geol. Tidsskrift.*, **6**, 142-194.

Ewing, J., Ewing, M., Aitken, T. and Ludwig, W. J. (1968), North Pacific sediment layers measured by seismic profiling. *Geophys. Monogr.*, **12**, 147-173.

Ewing, M. and Connary, S. D. (1970), Nepheloid layer in the North Pacific. *Geol. Soc. Am. Mem.*, **126**, 41-82.

_____ and Thorndike, E. M. (1965), Suspended matters in deep-ocean water. *Science,* **147**, 1291-1294.

Fairbridge, R. W. and Bourgeois, J. (eds.) (1978), *The Encyclopedia of Sedimentology.* Dowden, Hutchinson & Ross, Stroudsburg, Pa.

Faugères, J. C., Mézerais, M. L. and Stow, D. A. V. (1993), Contourite drift types and their distribution in the North and South Atlantic Ocean basins. *Sed. Geol.*, **82**, 189-203.

Feely, R. A. (1976), Evidence for aggregate formation on a nepheloid layer and its possible role in the sedimentation of particulate matter. *Marine Geol.*, **20**, M7-M13.

Fisk, H. N., McFarlan, E., Jr., Kolb, C. R. and Wilbert, L. J., Jr. (1954), Sedimentary framework of the modern Mississippi delta. *J. Sed. Petrol.*, **24**, 76-99.

Flemming, B. W. and Bartholomä, A. (eds.) (1995), Tidal signatures in modern and ancient sediments. *Int. Assoc. Sediment. Spec. Publ.*, (24).

Flint, S. S., Aitken, J. E. and Hampson, G. (1995), Application of sequence stratigraphy to coal-bearing coastal plain successions: implications for the UK coal measures. *In*: Whateley, M. K. G. and Spears, D. A. (eds.), *European Coal Geology. Geol. Soc. Spec. Publ.*, **82**, 1-16.

Folk, R. L. (1951), Stages of textural maturity in sedimentary rocks. *J. Sed. Petrol.*, **21**, 127-130.

_____ (1954), The distinction between grain size and mineral composition in sedimentary rock nomenclature. *J. Geol.*, **62**, 344-359.

_____ (1955), Student operator error in determination of roundness, sphericity, and grain size. *J. Sed. Petrol.*, **25**, 297-301.

_____ (1965a), *Petrology of Sedimentary Rocks.* Hemphill's, Austin.

_____ (1965b), Henry Clifton Sorby (1826-1908), the founder of petrography. *J. Geol. Educ.*, **13** (2), 43-47.

_____ (1966), A review of grain-size parameters. *Sedimentology*, **6**, 73-93.

_____ (1968), *Petrology of Sedimentary Rocks.* Hemphill's, Austin.

_____ and Ward, W. C. (1957), Brazos River bar: A study in the significance of grain size parameters. *J. Sed. Petrol.*, **27**, 3-26.

Forel, F. A. (1887), Le ravin sous-lacustre du Rhône dans le lac Léman. *Soc. Vaudoise Sci. Nat.*

*Bull.*, **23**, 85-107.
Frank, M. (1936), Der Faziescharakter der Schietgrenzen der süddeutschen und kalkalpinen Trias. *Zentralbl. Mineral. Geol. Paläont., Abt. B*, 475-502.
Friedman, G. M. (1958), Determination of sieve-size distribution from thin-section data for sedimentary petrological studies. *J. Geol.*, **66**, 394-416.
_____ (1962), On sorting, sorting coefficients and the lognormality of the grain-size distribution of sandstones. *J. Geol.*, **70**, 737-753.
_____ and Johnson, K. G. (1982), *Exercises in Sedimentology*. John Wiley, New York.
_____ and Sanders, J. E. (1978), *Principles of Sedimentology*. John Wiley, New York.
Fritz, W. J. and Moore, J. N. (1988), *Basics of Physical Stratigraphy and Sedimentology.* John Wiley, New York.
Fujii, K. (1956), Sandstones of the Mesozoic formations in the Yatsushiro districts, Kumamoto Prefecture, Kyushu, Japan. *J. Geol. Soc, Japan*, **62**, 193-211. (J+E)
Gale, A. S. (1998), Cyclostratigraphy. *In*: Doyle, P. and Bennett, M. R. (eds.), *Unlocking the Stratigraphical Record. Advances in Modern Stratigraphy.* John Wiley & Sons, Chichester, 195-220.
Ganly, P. (1856), Observations on the structure of strata. *J. Geol. Soc. Dublin*, 7, 164-166.
Gao, Z. H. and Eriksson, K. A. (1991), Internal tide deposits in an Ordovician submarine channel: previously unrecognized facies? *Geology*, **19** (7), 734-737.
_____ Eriksson, K. A., He, Y. B., Luo, S. S. and Guo, J. H. (1998), *Deep-Water Traction Current Deposits – A Study of Internal Tides, Internal Waves, Contour Currents and Their Deposits.* Science Press, Beijing.
Gardner, W. D. and Sullivan, L. G. (1981), Benthic storms: temporal variability in a deep-ocean nepheloid layer. *Science*, **213**, 329-331.
Gilbert, G. K. (1885), The topographic features of lake shores. *Annual Report of the US Geological Survey*, **5**, 75-123.
Gilligan, A. (1920), The petrography of the Millstone Grit of Yorkshire. *Quart. J. Geol. Soc. London*, **75**, 260-262.
Ginsburg, R. (ed.)(1973), *Evolving Concepts in Sedimentology*. Johns Hopkins Univ. Press, Baltimore.
Goldman, M. I. (1915), Petrographic evidence on the origin of the Catahoula sandstone of Texas. *Am. J. Sci., Ser. 4*, **39**, 261-287.
_____ (1950), What is sedimentology? *J. Sed. Petrol.*, **20** (2), 118.
Goldich, S. S. (1938), A study in rock weathering. *J. Geol.*, **46**, 17-58.
Gould, H. R. (1951), Some quantitative aspects of Lake Mead turbidity currents. *SEPM Spec. Publ.*, **2**, 34-52.
Grabau, A. W. (1913), *Principles of Stratigraphy*. A. G. Seiler.
Graton, L. C. and Fraser, H. J. (1935), Systematic packing of spheres, with particular relation to porosity and permiability. *J. Geol.*, **43**, 785-909.
Greensmith, J. T., 1976: *Petrology of the Sedimentary Rocks. 7th Ed.* Unwin Hyman, London.
Gressly, A., 1838, Observations géologiques sur le Jura Soleurois. *Nouv. Mém. Soc. Helv. Sci. Natur.*, **2**.
Grover, N. C. and Howard, C. S. (1937), The passage of turbid water through Lake Mead [Arizona and Nevada]. *Am. Soc. Civil Eng., Proc.*, **63** (1), 643-655.
Groves, A. W., 1931: The unroofing of the Dartmoor granite and distribution of its detritus in southern England. *Quart. J. Geol. Soc. London*, **87**, 62-96.
Guex, J. (1991), *Biochronological Correlation*. Springer-Verlag, Berlin.
Hadding, A. (1929), The pre-Quaternary sedimentary rocks of Sweden. III. The Paleozoic and Mesozoic sandstones of Sweden. *Lunds Univ. Årsskr., N.F., Avd. 2*, **25**.
Haeckel, E. (1887), Report on the Radiolaria collected by HMS *Challenger* during the years 1873-1876. *Rept. Sci. Results H.M.S. Challenger, Zoology*, **18**.

Hager, B. H., Clayton, R. W., Richards, M. A., Comer, R. P. and Dziewonski, A. M. (1985), Lower mantle heterogeneity, dynamic topography and the geoid. *Nature*, **313**, 541-545.

Hahn, H. H. and Stamm, W. (1970), The role of coagulation in natural waters. *Am. J. Sci.*, **768**, 354-368.

Hailwood, E. A. (1989), Magnetostratigraphy. *Geol. Soc. Spec. Rep.*, **19**.

Hall, J. (1843), Remarks upon casts of mud furrows, wave lines, and other markings upon rocks of the New York system. *Assoc. Am. Geol. Rep. for 1843*, 422-432.

_____ (1859), Description and figures of the organic remains of the lower Helderberg Group and the Oriskany Sandstone. *Natural History of New York. Part 6. Palaeontology. Geol. Surv. Albany, New York*, **3**, 532p.

Hallam, A. (1981), *Facies Interpretation and the Stratigraphic Record.* Freeman, Oxford.

Hampton, M. A. (1972), The role of subaqueous debris flows in generating turbidity currents. *J. Sed. Petrol.*, **42**, 775-793.

Hansen, E. (1971), *Strain Facies.* Springer-Verlag, Berlin.

Haq, B. U., Hardenbol, J. and Vail, P. R. (1987), Chronology of fluctuating sea levels since the Triassic. *Science*, **235**, 1156-1167.

Hara, I., Paulitsch, P. and Hide, K. (1976), An estimation method of velocity of orogenic movement. *Abh. N. Jb. Geol. Paläont.*, **151**, 58-72.

Harada, K. (1999), *Basics of Physical Stratigraphy and Sedimentology* (by W. J. Fritz and J. N. Moore) (translated into Japanese). Aichi Shuppan, Tokyo. (J)

Harada, M. (1995), Minamata disease: methylmercury poisoning in Japan caused by environmental pollution. *Crit. Rev. Toxicol.*, **25**, 1-24.

Harata, T. (1965), Some directional structures in the flysch-like beds of the Shimanto Terrain in the Kii Peninsula, Southwest Japan. *Mem. Coll. Sci., Univ. Kyoto, Ser. B*, **32** (2), 103-176.

Harnois, L. (1988), The CIW index: A new chemical index of weathering. *Sediment. Geol.*, **55**, 319-322.

Hasegawa, T. (1995), Correlation of the Cenomanian/Turonian boundary between Japan and Western Interior of the United States. *J. Geol. Soc. Japan*, **101** (1), 2-12.

Hashimoto, M. (ed.) (1991), *Geology of Japan.* Kluwer Academic Publishers, Dordrecht.

_____ and Uyeda, S. (eds.) (1983), *Accretion Tectonics in the Circum-Pacific Regions.* Terra Scientific Publishing Co., Tokyo.

Hatch, F. H. and Rastall, R. H. (1913), *The Petrology of the Sedimentary Rocks.* George Allen, London.

Hattori, M. (1967), Recent sediments of Sendai Bay, Miyagi Prefecture, Japan. *Sci. Rep. Tohoku Univ., 2nd Ser.*, **39** (1), 1-61.

Haug, Émile (1900), Les géosynclinaux et les aires continentales. *Soc. Géol. France Bull., 3me ser.*, **28**, 617-711.

Hayashi, T. (1966), Clastic dikes in Japan (1). *Japan. J. Geol. Geogr.*, **37** (1), 1-20.

Heckel, P. H. (1972), Recognition of ancient shallow marine environments. *Soc. Econ. Paleont. Mineral. Spec. Publ.*, **16**, 226-286.

Heezen, B. C. (1960), Turbidity currents. *In: McGraw-Hill Encyclopedia of Science and Technology*, McGraw-Hill, 146-147.

_____ and Ewing, M. (1952), Turbidity currents and submarine slumps, and the 1929 Grand Banks earthquake. *Am. J. Sci.*, **250**, 849-873.

_____ and Hollister, C. D. (1964), Deep-sea current evidence from abyssal sediments. *Mar. Geol.*, **1**, 141-174.

_____ _____ (1971), *The Face of the Deep.* Oxford Univ. Press, New York.

_____ _____ and Ruddiman, W. F. (1966), Shaping of the continental rise by deep geostrophic contour currents. *Science*, **152**, 502-508.

_____ MacGregor, I. D., Foreman, H. P., Forestall, G., Hekel, H., Hesse, R., Hoskins, R. H.,

Jones, E. J. W., Krasheninnikov, V. A., Okada, H. and Ruef, M. H. (1973), Diachronous deposits: a kinematic interpretation of the post-Jurassic sedimentary sequence on the Pacific plate. *Nature*, **241** (5384), 25-32.

Helmbold, R. and van Houten, F.B. (1958), Contribution to the petrography of the Tanner Graywacke. *Geol. Soc. Am. Bull.*, **69**, 301-314.

Hess, H. H. (1962), History of ocean basins. *In*: Engel, A. E. J. *et al.* (eds.), *Petrologic Studies: A Volume in Honor of A.F. Buddington*. Geol. Soc. Am., 599-620.

Hesse, R., Foreman, H. P., Forestall, G., Heezen, B. C., Hekel, H., Hoskins, R. H., Jones, E. J. W., Kaneps, A. G., Krasheninnikov, V. A., MacGregor, I. D., and Okada, H. (1974), Walther's facies rule in pelagic realm – a large-scale example from the Mesozoic-Cenozoic Pacific. *Z. Deutsch. Geol. Ges.*, **123**, 151-172.

Higham, N. (1963), *A Very Scientific Gentleman. The Major Achievements of Henry Clifton Sorby*. Pergamon Press, Oxford.

Hill, M. N. (ed.) (1963), *The Sea. Vol. 3*. Wiley-Interscience, New York.

Hirano, H., Nakayama, E. and Hanano, S. (1991), Oceanic Anoxic Event at the boundary of Cenomanian/Turonian ages – First report on the Cretaceous Yezo Supergroup, Hokkaido, Japan from the view of biostratigraphy of megafossils, sedimentary- and ichno-facies and geochemistry. *Bull. Sci. Eng. Res. Lab., Waseda Univ.*, (131), 52-59. (J+E)

_____ Toshimitsu, S., Matsumoto, T. and Takahashi, K. (2000), Changes in Cretaceous ammonoid diversity and marine environments of the Japanese Islands. *In*: Okada, H. and Mateer, N.J . (eds.), *Cretaceous Environments of Asia*. Elsevier, Amesterdam, 145-154.

Hirayama, J., Ichikawa, T., Moritani, T. and Mizuno, A. (trans.) (1967), *N. M. Strakhov's Genesis of Sedimentary Rocks (I)*. K. K. Rachisu, Tokyo. (J)

_____ _____ _____ _____ (trans) (1968), *N. M. Strakhov's Genesis of Sedimentary Rocks (II)*. K. K. Rachisu, Tokyo. (J)

_____ _____ _____ _____ (trans) (1971), *N. M. Strakhov's Genesis of Sedimentary Rocks (III)*. K. K. Rachisu, Tokyo. (J)

Hirayama, J. and Nakajima, T. (1977), Analytical study of turbidites, Otadai Formation, Boso Peninsula, Japan. *Sedimentology*, **24**, 747-779.

_____ and Suzuki, Y. (1968), Analysis of layers: An example in flysch-type alternations. *Chiku Kagaku (Earth Sci.)*, **22** (2), 43-62. (J+E)

Hisada, K. and Arai, S. (1993), Detrital chrome spinels in the Cretaceous Sanchu sandstone, central Japan; indicator of serpentinite protrusion into a fore-arc region. *Palaeogeography Palaeoclimatology Palaeoecology*, **105**, 95-109.

_____ _____ and Lee, Y. I. (1999), Tectonic implication of Lower Cretaceous chromian spinel-bearing sandstones in Japan and Korea. *The Island Arc*, **8**, 336-348.

Hisatomi, K. (1984), Sedimentary environment and basin analyses of the Miocene Kumano Group in the Kii Peninsula, Southwest Japan. *Mem. Fac. Sci., Kyoto Univ.*, **50** (1-2), 1-65.

Hjulström, F. (1939), Transportation of detritus by moving water. *In*: Trask, P. D. (ed.), *Recent Marine Sediments*. Am. Assoc. Petrol. Geol.,Tulsa, 5-31.

Hole, M. J. (1998), Stratigraphical applications of radiogenic geochemistry. *In*: Doyle, P. and Bennett, M. R. (eds.), *Unlocking the Stratigraphical Record. Advances in Modern Stratigraphy.* John Wiley & Sons, Chichester.

Hollister, C. D. (1993), The concept of deep-sea contourites. *Sediment. Geol.*, **82** (1-4), 5-11.

_____ and McCave, I. N. (1984), Sedimentation under deep-sea storms. *Nature*, **309** (5965), 220-225.

_____ Johnson, D. A. and Lonsdale, P. F. (1974), Current-controlled abyssal sedimentation: Samoan passage, Equatorial West Pacific. *J. Geol.*, **82**, 275-300.

Holz, M., Kalkreuth, W. and Banerjee, I. (2002), Sequence stratigraphy of paralic coal-bearing strata: an overview. *Intern. J. Coal Geol.*, **48**, 147-179.

Horsfield, B. and Stone, P. B. (1972), *The Great Ocean Business*. Hodder and Stoughton, London.
Hough, J. L. (1950), Editorial note. *J. Sed. Petrol.*, **20** (2), 118-119.
Howland, A.L. ((1975), William C. Krumbein: The making of a methodologist. *Geol. Soc. Amer.*, Mem. 42, xi-xviii.
Hoyanagi, K. (1989), Progradational lithofacies change of turbidite sequence, Middle Miocene Kotambetsu Formation, central Hokkaido, Japan. *J. Geol. Soc. Japan*, **95** (7), 509-525. (J+E)
_____ (1992), Sedimentary environments and sequence stratigraphy of the Neogene System in the Haboro area, northern part of the central Hokkaido, Japan. *Mem. Geol. Soc. Japan*, (37), 227-238. (J+E)
Hsü, K. J. (1972a), Origin of saline giants: a critical review after the discovery of the Mediterranean Evaporite. *Earth-Science Review*, **8**, 371-396.
_____ (1972b), When the Mediterranean dried up. *Scientific American*, **227** (6), 27-36.
_____ (1978), When the Black Sea was drained. *Scientific American*, **238**, 53-63.
_____ (1983), *The Mediterranean was a Desert. A Voyage of the Glomar Challenger*. Princeton Univ. Press, Princeton.
_____ (1989), *Physical Principles of Sedimentology*. Springer-Verlag, Berlin.
_____ (1992), *Challenger at Sea. A Ship That Revolutionized Earth Science*. Princeton Univ. Press, Princeton.
_____ (2004), *Physics of Sedimentology. 2nd Ed.* Springer-Verlag, Berlin.
_____ Cita, M. B. and Ryan, W. B. F. (1973), The origin of the Mediterranean evaporites. *Init. Repts. DSDP*, **13**, 1203-1231.
_____ Montadert, L., Bernoulli, D., Cita, M. B., Erickson, A., Garrison, R. F., Kidd, R. B., Mèlierés, F., Müller, C. and Wright, R. (1977), History of the Mediterranean salinity crisis. *Nature*, **267**, 399-403.
Hubert, J. F. (1962), A zircon-tourmaline-rutile maturity index and the interdependence of the composition of heavy mineral assemblages with the gross composition and texture of sandstones. *J. Sed. Petrol.*, **32**, 440-450.
_____ (1964), Textural evidence for the deposition of many western North Atlantic deep sea sands by ocean-bottom currents rather than turbidity currents. *J. Geol.*, **72**, 757-785.
Huckenholz, H. G. (1963), Mineral composition and texture in graywackes from the Harz Mountains (Germany) and in arkoses from the Auvergne (France). *J. Sed. Petrol.*, **33**, 914-918.
Hull, E. (1862), On iso-diametric lines, as means of representing the distribution of sedimentary clay and sandy strata, as distinguished from calcareous strata, with special reference to the Carboniferous rocks of Britain. Quart. *J. Geol. Soc. London*, **18**, 127-146.
Hussong, D. M., Uyeda, S. *et al.* (1982), Site 452: Mesozoic Pacific Ocean Basin. *Init. Repts. DSDP*, **60**, 77-99.
Hutton, J. (1788), Theory of the Earth. *Trans. Roy. Soc. Edinburgh*, **1** (2), 209-304.
_____ (1795), *Theory of the Earth with Proofs and Illustrations*. 2 vols. William Creech, Edinburgh.
Hynek, B. M. and Phillips, R. J. (2001), Evidence for extensive denudation of the Martian highlands. *Geology*, **29** (5), 407-410.
Ichihara, T., Takatsuka, K. and Shimoyama, S. (1996), 'Ichnostratigraphy' – the use of ichnofacies in stratigraphy – *J. Geol. Soc. Japan*, **102**, 685-699. (J+E)
Iijima, A. (1959), On relationship between the provenances and the depositional basins, considered from the heavy mineral associations of the Upper Cretaceous and Tertiary formations in central and southern Hokkaido, Japan. *J. Fac. Sci., Univ. Tokyo, Sec. II*, **11** (4), 339-385.
_____ (1978), Shallow-sea, organic origin of the Triassic bedded chert in central Japan. *J. Fac. Sci., Univ. Tokyo, Sec. II*, **19** (5), 369-400.

_____ and Utada, M. (1966), Zeolites in sedimentary rocks, with reference to the depositional environments and zonal distribution. Sedimentology, 7, 327-357.

Ijiri, S. (1950), Fundamental problems in sedimentology. *Kagaku (Science)*, **20** (7), 298-302. (J)

Ikebe, Y. (1962), Tectonic developments of oil-bearing Tertiary and migrations of oil, in Akita oil fields, Japan. *Rep. Res. Insti. Underground Resources, Mining College, Akita Univ.*, (26), 1-59. (J+E)

Ikehara, K. (1991), Modern sedimentation off San'in district in the southern Japan Sea. *In*: Takano, K. (ed.), *Oceanography in Asian Marginal Seas*. Elsevier, Amsterdam, 143-161.

Ikingura, J. R. and Akagi, H. (1996), Monitoring of fish and human exposure to mercury due to gold mining in the Lake Victoria gold-fields, Tanzania. *Sci. Total Environ.*, **191**, 59-68.

Illing, L. V. (1954), Bahama calcareous sands. *Bull. Am. Assoc. Petrol. Geol.*, **38**, 1-95.

Imai, H. (1921), On the relationship between the Horonai Formation and coal-bearing formations in the Ishikari coal-field. *Tohoku Imp. Univ., Inst. Geol. Pal. Contr.*, (1), 1-37. (J)

_____ (1931), *Chiso-gaku (Lithology)*. Kokon Shoin, Tokyo. (J)

Imai, I. (1966), *The Dawning of Japanese Geology*. Rateisu, Tokyo. (J)

Ingersoll, R. V. (1983), Petrofacies and provenance of Late Mesozoic forearc basin, northern and central California. *Am. Assoc. Petrol. Geol. Bull.*, **67**, 1125-1142.

Ingle, J. C. and Kariq, D. E. *et al.* (1975), Site 298. *Initial Reports of the Deep Sea Drilling Project*, Washington DC, **31**, 37-350.

Inman, D. L. (1952), Measures for describing the size distribution of sediments. *J. Sed. Petrol.*, **22**, 125-145.

Inouchi, Y., Tokuoka, T., Takayasu, H., Anma, K., Makino, Y. and Nirei, H. (eds.) (1990), Lakes: Origin, environment and geology. *Mem. Geol. Soc. Japan*, (36), 1-61. (J)

Isozaki, Y. (1993), Superanoxia and supercontinent: An example of P-T boundary recorded in deep-sea sediments. *Earth Monthly*, **15**, 311-313. (J)

_____ (1997), Permo-Triassic boundary superanoxia and stratified superocean: Records from a lost deep-sea. *Science*, **276**, 235-238.

Ito, M. (1985), The Ashigara Group: a progressive submarine fan-fan delta sequence in a Quaternary collision boundary, north of Izu Peninsula, central Honshu, Japan. *Sediment. Geol.*, **45**, 261-292.

_____ (1992), Sequence stratigraphic interpretation of methane gas production in the Kazusa Group, a Plio-Pleistocene forearc basin fill in the Boso Peninsula, Japan. *J. Coll. Arts Sci., Chiba Univ.*, **B-25**, 107-114.

_____ (1999), A fundamental framework of sequence stratigraphy. *J. Geol. Soc Japan*, **105** (7), 508-520. (J+E)

Jacobs, M. B. (1978), Nepheloid sediments and nephelometry. *In* Fairbridge, R. W. and Bourgeois, J. (eds.), *The Encyclopedia of Sedimentology*. Dowden, Hutchinson & Ross, Stroudsburg, Pa., 495-498.

_____ Thorndike, E. M. and Ewing, M. (1973), A comparison of suspended particulate matter from nepheloid and clear water. *Mar. Geol.*, **14**, 117-128.

Jakosky, B. (2000), Unearthing seeps and springs on Mars. *Planetary Report*, **20** (5), 12-16.

Jameson, R. (1808), *Elements of Geognosy*. Reprint by Jameson, R. and White, G. W. (1976), Hafner, New York.

Jenkyns, H. C. (1976), Sediments and sedimentary history of the Manihiki Plateau, South Pacific Ocean. *Init. Repts. DSDP*, **33**, 873-890.

_____ (1980), Cretaceous anoxic events: from continents to oceans. *J. Geol. Soc. London*, **137**, 171-188.

Jervey, M. T. (1988), Quantitative geological modelling of siliciclastic rock sequences and their seismic expressions. *In*: Wilgus, C. K. *et al.* (eds.), *Sea Level Changes: An Integrated Approach*. SEPM Special Publ., **42**, 47-69.

Johnson, D. A. (1972), Ocean floor erosion in the equatorial Pacific. *Geol. Soc. Am. Bull.*, **83**, 3121-3244.

Johnson, D.W. (1939), *The origin of submarine canyons. A critical review of hypotheses.* Columbia Univ. Press, New York.

Jones, E. J. W. (1999), *Marine Geophysics.* Wiley, Chichester.

Jones, O. T. (1938), On the evolution of a geosyncline. *Quart. J. Geol. Soc. London*, **94**, lx-cx.

Judd, J. W. (1908), Henry Clifton Sorby, and the birth of microscopical petrology. *Geol. Mag., New Series V*, **5** (5), 193-204.

Jukes-Browne, A. J. (1992), *The Building of the Brtish Isles.* Stanford, London.

Kagami, H., Shiono, K. and Taira, A. (1983), Subduction and forming processes of an accretionary prism in the Nankai Trough. *Kagaku (Science)*, **53**, 429-438. (J)

KAIKO I Research Group (1985), *Detailed Topography of the Trenches and Troughs around Japan.* Univ. Tokyo Press, Tokyo.

_____ (1986), *Topography and Structure of Trenches around Japan – Data Atlas of Franco-Japanese KAIKO Project, Phase I.* Univ. Tokyo Press, Tokyo.

KAIKO II Research Group (eds.) (1987), *6000 Meters Deep: A Trip to the Japanese Trenches. Photographic Records of the Nautile Dives in the Japanese Subduction Zones.* Univ. Tokyo Press, Ifremer, and CNRS. (J+E)

Kamada, Y. (1979), Nature and development of shelf deposit around Kyushu Island. *Report of the 8th Regional Symposium, Seikai National Fisheries Research Institute, Fisheries Agency, Japan*, 37-60. (J)

Kanmera, K. (1969), Litho- and bio-facies of Permo-Triassic geosynclinal limestone of the Sambosan Belt in southern Kysuhu. *Palaeont. Soc. Japan Spec. Publ.*, (14), 13-39.

_____ (1976), Relationship between ancient and modern geosynclinal sediments (I), (II). *Kagaku (Science)*, **46** (5), 287-291; (6), 371-378. (J)

_____ (1977), General aspects and recognition of olistostromes in the geosynclinal sequence. *Monogr. Assoc. Geol. Collab. Japan*, (20), 145-159. (J+E)

_____ Mizutani, S. and Chinzei, K. (eds.) (1979), *Materials and Environments of the Earth's Surface. Iwanami's Earth Science Series 5*, Iwanami Shoten, Tokyo.

Kato, Y. (1997), The tectonic landform in the continental slope deformed by the collision of the Amami Plateau. *J. Geogr.*, **106** (4), 567-579. (J+E)

Katsura, T. (1992), Submarine geology in the vicinity of Tsushima – Goto Retto region. *Rep. Hydrographic Res.*, (28), 55-138. (J+E)

Katto, J. (1960), Some problematica from the so-called unknown Mesozoic strata of the southern part of Shikoku, Japan. *Sci. Rep. Tohoku Univ., 2nd Ser., Spec. Vol. (Hanzawa Memorial Volume)*, (4), 323-334.

_____ (1961), Sedimentary structures from the Shimanto terrain, Shikoku, Southwest Japan. *Res. Rep. Kochi Univ., Nat. Sci.*, **10** (6), 135-142.

Kay, M. (1951), North American geosynclines. *Mem. Geol. Soc. Am.*, **48**.

Kazanskii, Yu.P. (1976), *Sedimentologia.* Nauka, Novosibirsk. (in Russian)

Kelling, G. (1964), The turbidite concept in Britain. *In*: Bouma, A. H. and Brouwer, A. (eds.), *Turbidites.* Elsevier, Amsterdam, 75-92.

Kelley, S. and Bluck, B. J. (1989), Detrital mineral ages from the Southern Uplands using $^{40}$Ar-$^{39}$Ar laser probe. *J. Geol. Soc. London*, **146**, 401-403.

Kennett, J. P., Burns, R. E., Andrews, J. E., Churkin, M., Jr., Davies, T. A., Dumitrica, P., Edwards, A. R., Galehouse, J. S., Packham, G. H. and van der Lingen, G. J. (1972), Australian-Antarctic continental drift, palaeocirculation changes and Oligocene deep-sea erosion. *Nature Physical Science*, **239**, 51-55.

Kiminami, K. (1975), Sedimentology of the Nemuro Group (Part I). *J. Geol. Soc. Japan*, **81** (4), 215-232.

_____ (1975–76), Sedimentology of the Nemuro Group (Part II-IV). *J. Geol. Soc. Japan*, **81** (11), 697-708; (12), 755-768; **82** (12), 773-782. (J+E)

_____ Kumon, F., Nishimura, T. and Shiki, T. (1992), Chemical composition of sandstones derived from magmatic arcs. *Mem. Geol. Soc. Japan*, (38), 361-372.

_____ _____ Miyamoto, T., Suzuki, S., Takeuchi, M. and Yoshida, K.(2000), Revised BI diagram and provenance types of the Permian-Cretaceous sandstones in the Japanese Islands. *Mem. Geol. Soc. Japan*, (57), 9-18. (J+E)

Kimiya, K. (1975a), Tensile strength as a physical scale of weathering in granitic rocks. *J. Geol. Soc. Japan*, **81** (6), 349-364. (J+E)

_____ (1975b), Rate of weathering of gravels of granitic rocks in Mikawa and Tomikusa areas. *J. Geol. Soc. Japan*, **81** (11), 683-696. (J+E)

Kimura, G. (1997), Cretaceous episodic growth of the Japanese Islands. *The Island Arc*, **6**, 52-68.

Kimura, H. (1953-56), A fundamental study of sedimentation (Part 1-7). *J. Geol. Soc. Japan*, **59**, 533-543; **60**, 81-93; 228-240; 337-348; 505-516; **61**, 103-116; **62**, 472-489. (J+E)

Kimura, T. (1966), Thickness distribution of sandstone beds and cyclic sedimentations in the turbidite sequences at two localities in Japan. *Bull. Earthquake Res. Inst., Univ. Tokyo,* **44**, 561-607.

Kinoshita, K. (1962), Geosyncline and oil fields – with special reference to the Neogene oil fields of Japan. *J. Japan. Assoc. Petrol. Technol.*, **27**(6), 211-225. (J+E)

_____ (1973), *Science of Petroleum Resources*. Kyoritsu Pub. Co., Tokyo.(J)

Klein, G. deV. (1977), *Clastic Tidal Facies*. Continuing Education Publication Co., Champaign, Illinois.

Kobayashi, T. (1941), The Sakawa orogenic cycle and its bearing on the origin of the Japanese Islands. *J. Fac. Sci., Imp. Univ. Tokyo, Ser. 2*, **5** (7), 219-578.

_____ (1951), *Introduction to the Regional Geology of Japan – Origin of Japan and the Sakawa Cycle*. Asakura Shoten, Tokyo. (J)

_____ (1956), The Triassic Akiyoshi Orogeny. *Geotecton. Symp. zu Ehren von Hans Stille*, Ferd. Enke, Stuttgart.

Kober, L. (1933), *Die Orogentheorie*. Gebrüder Bornträger, Berlin.

Kolla, V., Moore, D. V. and Curray, J. R. (1976), Recent bottom-current activity in the deep western Bay of Bengal. *Mar. Geol.*, **21**, 255-270.

Komar, P. D. (1971), Hydraulic jumps in turbidity currents. *Geol. Soc. Am. Bull.,* **82**, 1477-1488.

Konishi, K., Niguchi, K. and Ohara, H. (1985), Holocene environments of the Okinawa sea areas based on raised reef drilling core samples. *In*: Kajiura, K. (ed.), *Ocean Characteristics and their Changes*. Koseisha-Koseikaku, Tokyo, 430-437. (J)

Köster, E. (1964), *Granulometrische und morphometrische Messmethoden an Mineral-körnen, Steinen und sonstigen Stoffen*. F. Enke Verlag, Stuttgart.

Koster, E. H. and Steel, R. J. (eds.) (1984), *Sedimentology of Gravels and Conglomerates. Canad. Soc. Petrol. Geol. Bull.*, **10**.

Kotake, N. (1989), Paleoecology of the *Zoophycos* producers. *Lethaia*, **22**, 327-341.

_____ (1994), Population paleoecology of the *Zoophycos*-producing animal. *Palaios*, **9**, 84-91.

Krinsley, D. (1962), Applications of electron microscopy to geology. *New York Acad. Sci. Trans.*, **25**, 3-22.

_____ and Doornkamp, J. (1973), *Atlas of Quartz Sand Surface Textures*. Cambridge Univ. Press, Cambridge.

_____ and Margolis, S. (1969), A study of quartz sand grain surface textures with the scanning electron microscope. *Trans. New York Acad. Sci., Ser. II,* **31** (5), 457-477.

_____ and Smalley, I. J. (1972), Sand. *Am. Sci.*, **60** (3), 286-291.

_____ and Smalley, I. J. (1973), The shape and nature of small sedimentary quartz particles. *Science*, **180**, 1277-1279.

_____ and Takahashi, T. (1962a), The surface textures of sand grains, an application of electron microscopy. *Science*, **138**, 923-925.

_____ _____ (1962b), The surface textures of sand grains, an application of electron microscopy. Glaciation. *Science*, **138**, 1262-1264.

_____ _____ (1962c), Application of electron microscopy to geology. *Trans. N.Y. Acad. Sci., Ser. II*, **25** (1), 3-22.

_____ _____ (1964), A technique for the study of surface textures of sand grains with electron microscopy. *J. Sed. Petrol.*, **34**, 423-426.

Krumbein, W. C. (1934), Size frequency distributions of sediments. *J. Sed. Petrol.*, **4**, 65-77.

_____ (1939), Preferred orientation of pebbles in sedimentary deposits. *J. Geol.*, **47**, 673-706.

_____ (1941a), Measurement and geological significance of shape and roundness of sedimentary particles. *J. Sed. Petrol.*, **11**, 64-72.

_____ (1941b), The effects of abrasion on the size, shape and roundness of rock fragments. *J. Geol.*, **49**, 482-520.

_____ and Pettijohn, F. J. (1938), *Manual of Sedimentary Petrography.* Appleton-Century Crofts, New York.

_____ and Sloss, L. L. (1953), *Stratigraphy and Sedimentation.* W. H. Freeman, San Francisco.

Krynine, P. D. (1941a), Differentiation of sediments during the life history of a landmass. *Geol. Soc. Am. Bull.*, **52**, 1915.

_____ (1941b), Graywacke and the petrology of the Bradford Oil Field. *Am. Assoc. Petrol. Geol. Bull.*, **25**, 2071-2074.

_____ (1941c), Paleogeography and tectonic significance of sedimentary quartzites. *Geol. Soc. Am. Bull.*, **52**, 1916.

_____ (1941d), Paleogeography and tectonic significance of graywacke. *Geol. Soc. Am. Bull.*, **52**, 1916.

_____ (1941e), Paleogeography and tectonic significance of arkoses. *Geol. Soc. Am. Bull.*, **52**, 1918-1919.

_____ (1942), Differential sedimentation and its products during one complete geosynclinal cycle. *First Pan-Am. Congr. Mining Eng. Geol.,* **2**, pt. 1, 536-561.

_____ (1946a), Microscopic morphology of quartz types. *Second Pan-Am. Congr. Mining Eng. Geol.,* **3**, 35-49.

_____ (1946b), The tourmaline group in sediments. *J. Geol.*, **54**, 65-87.

_____ (1948), The megascopic study and field classification of sedimentary rocks. *J. Geol.*, **56**, 130-165.

_____ (1951), Reservoir petrology of sandstones. *In*: Payne, T. G. *et a*l. (eds.), *Geology of Arctic Slope of Alaska.* U.S. Geol. Surv. Oil and Gas Inv. Map OM 126.

Ksiazkiewicz, M. (1960), Pre-orogenic sedimentation in the Carpathian Geosyncline. *Geol. Rundschau*, **50**, 8-31.

Kuenen, Ph.H. (1937), Experiments in connection with Daly's hypothesis on the formation of submarine canyons. *Leidsche Geol. Meded.*, **8**, 327-351.

_____ (1950a), Turbidity currents of high density. *Intl. Geol. Cogr., 18th, London, 1948, Rept. Pt. 8,* 44-52.

_____ (1950b), *Marine Geology.* John Wiley & Sons, Inc., New York.

_____ (1952), Estimated size of the Grand Banks turbidity current. *Am. J. Sci.*, **250**, 874-884.

_____ (1956), Experimental abrasion of pebbles: 2. Rolling by current. *J. Geol.*, **64**, 336-368.

_____ (1957), Sole markings of graded graywacke beds. *J. Geol.*, **65**, 231-258.

_____ (1958a), Experiments in geology. *Trans. Geol. Soc. Glasgow*, **23**, 1-9.

_____ (1958b), No geology without marine geology. *Geol. Rundschau*, 47, 1-10.

_____ (1963), Pivotability studies of sand by a shape-sorter. *Sedimentology*, **1**, 208-215.

_____ (1965), Value of experiments in geology. *Geol. Mijnbouw*, **44**, 22-36.

_____ and Migliorini, C. (1950), Turbidity currents as a cause of graded bedding. *J. Geol.*, **58**, 91-127.

Kuhn, T. S. (1962), *The Structure of Scientific Revolutions*. Univ. Chicago Press.

Kumon, F. and Kiminami, K. (1994), Modal and chemical compositions of the representative sandstones from the Japanese Islands and their tectonic implications. *Proc. 29th Intn. Geol. Congr., Part A*, 135-151.

_____ _____ Hoyanagi, K., Takeuchi, M., Musashino, M. and Miyamoto, T. (eds.) (2000), Compositions of siliciclastic rocks in relation to provenance, tectonics and sedimentary environments. *Mem. Geol. Soc. Japan*, (57), 1-240. (J+E, 3 papers in English)

Lane, M. D. (2004), Mars rocks! Deciphering minerals on Mars. *Planetary Report*, **24** (1), 12-17.

Langseth, M. G., Okada, H. *et al.* (1978), Near the Japan Trench transects begun. *Geotimes*, **23** (3), 22-26.

Larson, R. L. (1991a), Latest pulse of earth: evidence for a mid-Cretaceous superplume. *Geology*, **19**, 547-550.

_____ (1991b), Geological consequences of superplumes. *Geology*, **19**, 963-966.

_____ Fischer, A. G., Erba, E. and Premoli Silva, I. (1992), Summary of workshop results. *In*: Larson, R. L., Fischer, A. G., Erba, E. and Premoli Silva, I. (eds.), *Apticore–Albicore*. Joint Oceanographic Institutions, Inc., Washington, DC.

Lash, G. G. (1985), Recognition of trench fill in orogenic flysch sequences. *Geology*, **13**, 867-870.

Leeder, M.R. (1982), *Sedimentology. Process and Product*. Allen & Unwin.

_____ (1998), Lyell's *Principles of Geology*: foundations of sedimentology. *In*: Blundell, D. J. and Scott, A. C. (eds.), *Lyell: the Past is the Key to the Present. Geological Society, London, Spec. Publ.*, **143**, 97-110.

Leggett, J. K. (1985), Deep-sea pelagic sediments and palaeo-oceanography: a review of recent progress. *In*: Brenchley, P. J. and Williams, B. P. J. (eds.), *Sedimentology. Recent Developments and Applied Aspects*. The Geological Society, London, 95-121.

Leopold, L. B., Wolman, M. G. and Miller,J.P.(1964), *Fluvial Processes in Geomorphology*. Freeman, San Francisco, xiii + 522p.

Le Pichon, X., Kobayashi, K., Cadet, J. P., Iiyama, T., Nakamura,K., Pautot, G., Renard, V. and the Kaiko Scientific Crew (1987a), Project Kaiko – Introduction. *Earth Planet. Sci. Letts.*, **83**, *Special Issue*, 183-195.

_____ Iiyama, T., Boulègue, J., Charvet, J., Faure, M., Kno, K., Lallemant, S., Okada, H., Rangin, C., Taira, A., Urabe, T. and Uyeda, S. (1987b), Nankai Trough and Zenizu Ridge: a deep-sea submersible survey. *Earth Planet. Sci. Letts.*, **83**, *Special Issue*, 285-299.

Lewis, D. W. (1984), *Practical Sedimentology*. Hutchinson Ross, Stroudsburg.

Lindholm, R. C. (1987), *A Practical Approach to Sedimentology*. Allen & Unwin, London.

_____ (1972), Sedimentology of clastic rocks returned from the moon by Apollo 15. *Geol. Soc. Am. Bull.*, **83**, 2957-2970.

Lindsay, J. F. (1972), Transportation of detrital materials on the lunar surface. *Sedimentology*, **21**, 323-328.

_____ (1976), *Lunar Stratigraphy and Sedimentology*. Elsevier, Amsterdam.

_____ (1978), Lunar sedimentation. *In*: Fairbridge, R.W. and Bourgeois, J. (eds.), *The Encyclopedia of Sedimentology*. Dowden, Hutchinson & Ross, Stroudsburg, Pa., 458-461.

Linklater, E. (1972), *The Voyage of the Challenger*. Doubleday, Garden City.

Lombard, A. (1978), Sedimentology. *In*: Fairbridge, R.W. and Bourgeois, J. (eds.), *The Encyclopedia of Sedimentology*. Dowden, Hutchinson & Ross, Inc., Stroudsburg, Pa., 703-707.

Longman, C. D., Bluck, B. J. and van Breemen, O. (1979), Ordovician conglomerates and the evolution of the Midland Valley. *Nature*, **280**, 578-581.

Longwell, C. R. (ed.) (1949), *Sedimentary Facies in Geologic History*. Geol. Soc. Am. Mem., **39**.

Lonsdale, P. and Southard, J. B. (1974), Experimental erosion of North Pacific red clay. *Mar.*

*Geol.*, 17, M51-M60.

Lowe, D. R. (1975), Water escape structures in coarse-grained sediments. *Sedimentology*, **23**, 285-308.

Luyten, J. R. (1977), Scales of motion in the deep Gulf Stream and across the continental rise. *J. Mar. Res.*, **35**, 49-74.

Lyell, C., (1830-1833), *Principles of Geology*. Murray, London. 3 vols.

Lyman, B. S. (1877), *A General Report on the Geology of Yesso*. Hokkaido Colon. Dept., Sapporo.

_____ Munroe, H. S., Yamauchi, T., Akiyama, Y., Inagaki, T., Kuwada, T., Misawa, S., Takahashi, J., Kada, T., Ban, I., Saito, T., Shimada, J., Yamagiwa, E., Mayeda, S. and Nishiyama, S. (1876), *A Geological Sketch Map of the Island of Yesso, Japan. Scale: 1/2,000,000 of Nature.* Geol. Surv. Hokkaido, Japan. (J+E)

Mackie, Wm. (1896), The sands and sandstones of Eastern Moray. *Trans. Edinburgh Geol. Soc.*, 7, 148-172.

Maejima, W. (1986), Initial sedimentation and tectonics in the Cretaceous Yuasa-Aridagawa Basin, western Kii Peninsula, Southwest Japan: facies and genesis of the Yuasa Formation. *J. Geosci., Osaka City Univ.*, **29** (1), 1-44.

Maiklem, W. R. (1968), Some hydraulic properties of bioclastic carbonate grains. *Sedimentology*, **10**, 101-109.

Makino, Y. (1994), Wave ripple dynamics and the combined-flow modification of wave ripples in the intertidal zone of Ariake Bay (Kyushu, Japan). *Marine Geol.*, **120**, 63-74.

Makiyama, J. (1956), *Tectonic Geology*. Asakura Shoten. Tokyo. (J)

Malin, M. C. and Edgett, K. S. (2000), Evidence for recent groundwater seepage and surface runoff on Mars. *Science*, **288** (5475), 2330-2335.

Maruyama, S. (1993), Plume tectonics. *J. Geol. Soc. Japan*, **100** (1), 24-49.

Masuda, F. (1993), Sequence stratigraphy: Application to strata exposed on land in Japan. *J. Japan. Assoc. Petrol. Technol.*, **58** (4), 292-310. (J)

_____ (ed.) (2001), *Wave-generated Sedimentary Structures. Lecture Notes for Short Course No.1.* Sedimentological Society of Japan. (J)

Matsuda, H. (1995), Recognition of subaerial exposure surface in shallow-marine carbonate sequence based on stable carbon and oxygen isotopic composition. *J. Geol. Soc. Japan*, **101**, 889-901. (J+E)

_____ Tsuji, Y., Honda, N. and Saotome, J. (1995), Early diagenesis of Pleistocene carbonates from a hydrogeochemical point of view, Irabu Island, Ryukyu Islands: Porosity changes related to early carbonate diagenesis. *In*: Budd, D. A., Saller, A. H. and Harris, P. M. (eds.) (1995), *Unconformities and Porosity in Carbonate Strata*. Am. Assoc. Petrol. Geol. Bull., Mem., (63), 35-54.

Matsumoto, R. (1982), Environment and mechanism of deposition of radiolarian bedded chert: Geochemical approach. *Earth Monthly*, **42**, 527-535. (J)

_____ (1987a), Nature and occurrence of gas hydrates and their implications to geologic phenomena. *J. Geol. Soc. Japan*, **93**, 597-615. (J+E)

_____ (1987b), Gas hydrate and carbonate diagenesis. *J. Sed. Soc. Japan*, (27), 41-50. (J+E)

_____ (1991), Current topics of diagenesis. *J. Sed. Soc. Japan*, (34), 83-89.(J+E)

_____ (1995), Causes of the ä$^{13}$C anomalies of carbonates and a new paradigm "Gas Hydrate Hypothesis". *J. Geol. Soc. Japan*, **101**, 902-924. (J+E)

Matsumoto, T. (1936), Geology of the Onogawa basin, Kyushu. *J. Geol. Soc. Japan*, **43**, 758-786, 815-852. (J+E)

_____ (1938), Geology of the Goshora Island, Amakusa, with special reference to the Cretaceous stratigraphy. *J. Geol. Soc. Japan*, **45**, 1-47. (J+E)

_____ (1940), On the constituents of the uppermost Cretaceous Ryugase Group in Saghalien. *J. Geol. Soc. Japan*, 47 (564), 383-385. (J+E)

_____ (1949), *Assignments of Historical Geology of Japan*. Heibonsha, Tokyo. (J)

_____ (1967), Fundamental problems in the circum-Pacific orogenesis. *Tectonophysics,* **4** (4-6), 595-613.

Matsushita, H. (1949), Geology of the coal fields in northern Kyushu. *J. Mining Soc. Kyushu, Special Issue,* 1-57. (J)

McArthur, J. M. (1998), Strontium isotope stratigraphy. *In*: Doyle, P. and Bennett, M.R. (eds.), *Unlocking the Stratigraphical Record. Advances in Modern Stratigraphy.* John Wiley & Sons, Chichester, 221-341.

McBride, E. F. (1964), Review of turbidite studies in the United States. *In*: Bouma, A. H. and Brouwer, A. (eds.), *Turbidites*. Elsevier, Amsterdam, 93-105.

McIntyre, D. B. and McKirdy, A. (1997), *James Hutton. The Founder of Modern Geology*. Stationary Office, Edinburgh, 51p.

McKee, E. D. (1979), An introduction to the study of global sand seas. *In*: McKee, E. D. (ed.), *Global Sand Seas. U. S. Geol. Survey Prof. Paper*, **1052**, 1-19.

Menard, H. W. (1952), Deep ripple marks in the sea. *J. Sed. Petrol.*, **22**, 3-9.

Miall, A. D. (1978), Lithofacies types and vertical profile models in braided river deposits: a summary. *In*: Miall, A. D. (ed.), Fluvial sedimentology. *Canad. Soc. Petrol. Geol. Mem.*, **5**, 597-604.

_____ (1984), *Principles of Sedimentary Basin Analysis*. Springer-Verlag, New York.

_____ (1990), *Principles of Sedimentary Basin Analysis. 2nd Ed.* Springer-Verlag, New York.

_____ and Miall, C. E. (2001), Sequence stratigraphy as a scientific enterprise: the evolution and persistence of conflicting paradigms. *Earth-Science Reviews*, **54**, 321-348.

Middleton, G. V. (1965), Primary sedimentary structures and their hydrodaynamic interpretation. *Soc. Econ. Paleont. Mineral. Spec. Publ.*, **12**.

_____ (1966a), Small scale models of turbidity currents and the criterion for autosuspension. *J. Sed. Petrol.*, **36**, 202-208.

_____ (1966b), Experiments on density and turbidity currents. I. The motion of the head. *Canad. J. Earth Sci.*, **3**, 523-546.

_____ (1966c), Experiments on density and turbidity currents. II. Uniform flow of density currents. *Canad. J. Earth Sci.*, **3**, 627-637.

_____ (1967), Experiments on density and turbidity currents. III. Deposition of sediments. *Canad. J. Earth Sci.*, **4**, 475-505.

_____ 1973, Johannes Walther's law of the correlation of facies. *Geol. Soc. Amer. Bull.*, **84**, 979-988.

_____ 1978, Facies. *In*: Fairbridge, R. W. and Bourgeois, J. (eds.), *The Encyclopedia of Sedimentology*. Dowden, Hutchinson & Ross, Stroudsburg, Pa., 323-325.

_____ (2003a), Sedimentology, History. *In*: Middleton, G. V. (ed.), *Encyclopedia of Sediments and Sedimentary Rocks*. Kluwer Academic Pub., Dordrecht, 628-635.

_____ (ed.) (2003b), *Encyclopedia of Sediments and Sedimentary Rocks*. Kluwer Academic Pub., Dordrecht.

_____ and Hampton, M. A., (1973), Sediment gravity flows: Mechanics of flow and deposition. *In*: Middleton, G. V. and Bouma, A. H. (eds.), *Turbidites and Deep Water Sedimentation*. Short Course Notes. Pacific Sec., SEPM, 1-38.

_____ _____ 1976, Subaqueous sediment transport and deposition of sediment gravity flows. *In*: Stanley, D. J. and Swift, D. J. P. (eds.), *Marine Sediment Transport and Environmental Management*. Wiley, New York, 197-218.

Migliorini, C. (1943), Sul modo di formazione dei complessi tipo Macigno. *Boll. Soc. Geol. Ital.*, **62**, 48-50.

Mikami, T. (1969), A sedimentological study of the lower Permian Sakamotozawa Formation. *Mem. Fac. Sci., Kyushu Univ., Ser. D*, **19**, 331-372.

Miki, T. (1977), Graphitization of some Tertiary coals in Kyushu, Japan. *J. Geol. Soc. Japan*, **83** (9), 575-581. (J+E)
Milankovitch, M. (1941), *Kanon der Erdbestrahlung und seine Anwendung auf das Eis-eitenproblem.* Serbian Academy of Science, Belgrade, Special Publication 132, Section of Mathematical and Natural Sciences.
Miller, H. (1884), On boulder glaciation. *Proc. Roy. Phys. Soc. Edinburgh*, **8**, 156-189.
Milner, H. B. (1922), *An Introduction to Sedimentary Petrography.* Murby, London.
_____ (1929), *Sedimentary Petrography.* Murby, London.
_____ (1962), *Sedimentary Petrography. Vol. 2.* Allen & Unwin, London.
Milton, D. J. (1983), Water and processes of degradation in the martian landscape. *J. Geophys. Res.*, **78**, 4037-4047.
Minato, M. (1953), *Chiso-gaku (Lithology).* Iwanami Shoten, Tokyo. (J)
_____ (1973), *Chiso-gaku (Lithology). 2nd Ed.* Iwanami Shoten, Tokyo. (J)
Minoura, K. (1979), Sedimentological study of the Ryukyu Group. *Sci. Repts. Tohoku Univ., 2nd Ser.*, **48** (2), 1-69.
_____ and Hisajima, T. (1993), Evaporation of the palaeo-Mediterranean Sea. *Ocean Monthly*, **25** (6), 367-376. (J)
_____ and Nakata, T. (1994), Discovery of an ancient tsunami deposits in coastal sequences of southwest Japan: verification of a large historic tsunami. *The Island Arc*, **3**, 66-72.
_____ and Nakaya, S. (1991), Traces of tsunami preserved in inter-tidal lacustrine and marsh deposits: some examples from northeast Japan. *J. Geol.*, **99**, 265-287.
_____ _____ and Uchida, M. (1994), Tsunami deposits in a lacustrine sequence of the Sanriku coast, northeast Japan. *Sediment. Geol.*, **89**, 25-31.
Mitchell, A. H. G. and Reading, H. G. (1986), Sedimentation and tectonics. *In*: Reading, H. C. (ed.), *Sedimentary Environments and Facies.* Blackwell Scientific Publication, Oxford, 471-519.
Mitchum, R. M., Jr., Vail, P. R. and Thompson, S., III (1977), Seismic stratigraphy and global change of sea level. Part 2: The depositional sequence as a basic unit for stratigraphic analysis. *In*: Payton, C. E. (ed.), *Seismic Stratigraphy – Application to Hydrocarbon Exploration. Am. Assoc. Petrol. Geol. Mem.*, **26**, 53-62.
Mitsushio, H. (1967), Bottom sediments in bays of North Kyushu. *Mem. Fac. Sci., Kyushu Univ., Ser. D,* **18** (1), 7-34.
Mitsuzawa, K., Momma, H., Fukasawa, M. and Hotta, H. (1993), Observation of deep sea current and change of bottom shapes in the Suruga Trough. *IEEE Oceans '93 Proc.*, 149-154.
Miyamoto, T. (1980), Stratigraphical and sedimentological studies of the Cretaceous System in the Chichibu terrain of the Outer Zone of Southwest Japan. *J. Sci., Hiroshima Univ.*, (23), 1-128. (J+E)
Miyashiro, A. (1998), *What is Scientific Revolution?.* Iwanami Shoten, Tokyo. (J)
Miyata, Y. (1989), Verification of hydraulic equivalence during clastic sedimentation – Examples from tephra fallout and sandstones. *Mem. Fac. Sci., Kyushu Univ., Ser. D*, **26** (3), 243-263.
Mizuno, A. (1968), *Guide to the Study of Waters and Bottom Sediments.* K.K. Tatis, Tokyo. (J)
_____ Maruyama, S., Fujii, N., Yasuda, T., Noguchi, Y., Tsuru, S., Mochizuki, T., Ohshima, K., Yokota, S., Nakao, S. and Ono, M. (1970), Sedimentological study in the Miho Bay and near the Oki Islands, off the San'in coasts of southwestern Japan. *Bull. Geol. Surv. Japan*, **21** (4), 219-236. (J+E)
Mizutani, S. (1959), Clastic plagioclase in Permian greywacke from the Mugi area, Gifu Prefecture, central Japan. *J. Earth Sci., Nagoya Univ.*, 7, 108-136.
_____ (1966), Transformation of silica under hydrothermal conditions. *J. Earth Sci., Nagoya Univ.*, **14** (1), 56-88.
_____ (1970), Silica minerals in the early stage of diagenesis. *Sedimentology*, **15**, 419-436.
_____ (2000), On environmental analysis of sedimentary sequence on a basis of stochastic model.

*J. Social and Information Sci., Nihon Fukushi Univ.*, **3**, 1-20. (J+E)
_____ and Hattori, I. (1972), Stochastic analysis of bed-thickness distribution of sediments. *Math. Geol.*, **4**, 123-146.
_____ Saito, Y. and Kanmera, K. (eds.) (1987), *Sedimentary Rocks of Japan*. Iwanami Shoten, Tokyo. (J)
Moore, J. C. and Karig, D. E. (1976), Sedimentology, structural geology, and tectonics of the Shikoku subduction zone, southwestern Japan. *Geol. Soc. Am. Bull.*, **87**, 1259-1268.
Moore, R. C., 1949, Meaning of facies. *Geol. Soc. Am. Mem.*, **39**, 1-34.
Morton, A. C. (1985), A new approach to provenance studies: electron microprobe analysis of detrital garnets from Middle Jurassic sandstones of the northern North Sea. *Sedimentology*, **32**, 553-566.
Mulder, T. and Alexander, J. (2001), The physical character of subaqueous sedimentary density flows and their deposits. *Sedimentology*, **48**, 269-299.
Müller, H., 1933: Sedimentpetrographie und Geologie. *Z. Deut. Geol. Ges.*, **85**, 719-720.
Murakoshi, N. and Masuda, F. (1992), Estuarine, barrier-island to strand-plain sequence and related ravinement surface developed during the last interglacial in Paleo-Tokyo Bay, Japan. *Sedim. Geol.*, **80**, 167-184.
Murray, B. C.(1973), Mars from Mariner 9. *Scientific American*, **228** (1), 48-69.
_____ Ward, W. R. and Young, S. C. (1973), Periodic insulation variations on Mars. *Science*, **180**, 638-640.
Murray, J. and Renard, A. (1891), *Deep-Sea Deposits. Report on the Scientific Results of the Exploring Voyage of HMS Challenger, 1872-1876*. Longman, London.
Musashino, M. and Kasahara, K. (1986), Composition of detrital garnets from the Tamba Belt and the Ultra Tamba Zone. *Clastic Sediments (Journal of the Research Group of Clastic Sediments in Japan)*, (4), 41-47. (J+E)
Musser, G. (2004), The spirit of exploration. NASA's rover fights the curse of the Angry Red Planet. *Scientific American*, **290** (3). 52-57.
Mutch, T. A., Arvidson, R. E., Head, J. W., III, Jones, K. L. and Saunders, R. S. (1976), *Martian Sedimentation*. Princeton Univ. Press, New Jersey.
Muto, T. (1992), "Retreat" vs "equilibrium" theories for shoreline regression and transgression. *J. Sed. Soc. Japan*, (37), 9-19. (J)
_____ (1997), A note on the functions of accommodation and sediment supply. *J. Sed. Soc. Japan*, (46), 49-58. (J)
_____ and Steel, R. J. (1992), Retreat of the front in a prograding delta. *Geology*, **20**, 967-970.
Mutti, E. (1974), Examples of ancient and deep-sea fan deposits from circum-Mediterranean geosynclines. *Soc. Econ. Paleont. Mineral. Spec. Publ.*, **19**, 92-105.
_____ (1992), *Turbidite Sandstones*. Agip, Univ. Parma, Parma.
_____ and Ricci-Lucchi, F. (1972), Le torbiditi dell 'Appennino settentrionale: introduzione all' analisi di facies. *Mem. Soc. Geol. Ital.*, **11**, 161-199.
Nagahama, H. (1965), Diagonal bedding and accumulation of Tertiary sediments in northwestern Kyushu, Japan. *Rept. Geol. Surv. Japan*, (211), 1-66. (J+E)
Nagao, T. (1926–1928), *The Palaeogene Stratigraphy of Kyushu*. Reprinted in 1975 by Nagasaki Earth Science Association. (J)
Nanayama, F. and Nakagawa, M. (1995), Detrital chromian spinels from the Horobetsu-gawa Complex, in the Idon'nappu Belt, central Hokkaido, Japan. *J. Geol. Soc. Japan*, **101** (7), 549-552. (J)
Nara, M. (1997), High-resolution analytical method for event-sedimentation using *Rosselia socialis*. *Palaios*, **12**, 489-494.
Naruse, H., Tamura, T., Kubo, Y. and Masuda, F. (2001), *Sediment Gravity Flow Deposits and their Structures. Lecture Notes for Short Course No. 2*. Sedimentological Society of Japan. (J)

Nasu, N. (1961), Recent progress of marine geology in Japan. *In*: The Oceanographic Society of Japan (ed.), *Two Decades of the Oceanographic Society of Japan*, Oceanogr. Soc. Japan, 62-71. (J)
_____ (1981), Memorial to Prof. Takao Sakamoto. *J. Geogr.*, **90** (3), 220. (J)
Naumann, E. (1885), *Über den Bau und die Entstehung der japanischen Inseln*. Berlin.
Nemec, W. and Steel, R. J. (eds.) (1988), *Fan Deltas: Sedimentology and Tectonic Settings*. Blackie, Glasgow and London.
Nemoto, K. and Kroenke, L. W. (1981), Marine geology of the Hess Rise. 1. Bathymetry, surface sediment distribution, and environment of deposition. *J. Geophys. Res.*, **86** (B11), 10734-10752.
Neumayer, M. (1875), *Erdgeschichte, 1*. Bibliographisches Inst., Leipzig.
Nichols, G. (1999), *Sedimentology and Stratigraphy*. Blackwell, Oxford.
Niino, H. (1950), On the bottom sediments around a special submarine configuration. *Hydrographic Bull., Maritime Safety Agency, Tokyo, Special Number,* (7), 83-121. (J+E)
Niitsuma, N. (1971), Automatic grain-size analyzer for sedimentological investigation. *Tohoku Univ., Inst. Geol. Pal., Contr.*, (72), 25-36. (J+E)
Nilsen, T. H. (1985), *Modern and Ancient Alluvial Fan Deposits*. Van Nostrand Reinhold, New York.
Ninomiya, T., Ohmori, H., Hashimoto, K., Tsuruta, K. and Ekino, S. (1995), Expansion of methylmercury poisoning outside of Minamata: an epidemiological study on chronic methylmercury poisoning outside of Minamata. *Environ. Res.*, **70**, 47-50.
Nishiwaki, N. and Yamamoto, K. (1981), Mathematical geology and its application to sedimentology. *Earth Monthly*, **3** (6), 362-371. (J)
Noda, K. (1994), *Introduction to Study of Japanese Trace Fossils*. Rakushisha, Tokyo. (J)
Normark, W. R. (1970), Growth patterns of deep-sea fans. *Bull. Am. Assoc. Petrol. Geol.*, **54**, 2170-2195.
_____ (1978), Fan valleys, channels, and depositional lobes on modern submarine fans: Characters for recognition of sandy turbidite environments. *Am. Assoc. Petrol. Geol. Bull.*, **62**, 912-931.
Nota, D. J. G. (1974), In memorial Professor Dr. D. J. Doeglas. *Sedimentology*, **21**, 489-490.
Nriagu, J. O., Pfeiffer, W. C., Malm, O. and Mierle, G. (1992), Mercury pollution in Brazil. *Nature*, **356**, 389.
Oba, T. *et al.* (1991), Paleoenvironmental changes in the Japan Sea during the last 85,000 years. *Paleoceanography*, **6** (4), 499-518.
Ohara, J. (1961), Heavy mineral associations in the Paleogene System of some coal fields, north Kyushu, Japan. *Mem. Fac. Sci., Kyushu Univ., Ser. D (Geol.)*, **11**, 381-418.
O'Halloran, D., Green, C., Harley, M., Stanley, M. and Knill, J. (eds.)(1994), *Geological and Landscape Conservation. Proceedings of the Malvern International Conference 1993.* The Geological Society, London.
Ohshima, K., Inoue, E., Onodera, K., Yuasa, M. and Kuroda, K. (1982), Sediments of the Tsushima Strait and Goto-nada Sea, northwestern Kyushu. *Bull. Geol. Surv. Japan*, **33** (7), 321-350. (J+E)
_____ Onodera, K. and Arita, M. (1975), Dispersal of released crude oil and submarine sediments. *Chishitsu News, Geological Survey of Japan*, (254), 32-41. (J)
_____ and Saito, Yo. (1993), Environmental resources of Tokyo Bay. *Earth Sci.*, **47** (3), 213-231. (J+E)
Ohta, S. (1983), Photographic census of large-sized benthic organisms in the bathyal zone of Suruga Bay, central Japan. *Bull. Ocean Res. Inst., Univ. Tokyo*, **15**, 1-244.
_____ (1989), Current measurements with bottom-moored current meter systems in the axis of the Suruga Bay. *In*: Ohta, S. (ed.), *Preliminary Report of the Hakuho Maru Cruise KH-88-4.* Ocean Res. Inst., Univ. Tokyo, 63-70.

Okada, H. (1964), Serpentine sandstone from Hokkaido. *Mem. Fac. Sci., Kyushu Univ., Ser. D, Geol.,* **15** (1), 23-38.
_____ (1967), Some problems of sandy sediments in geosynclines. *Kagaku (Science),* **37** (5), 270-276. (J)
_____ (1968a), Classification and nomenclature of sandstones. *J. Geol. Soc. Japan,* 74 (7), 589-594. (J+E)
_____ (1968b), Inverted grading in gravels and conglomerates. *J. Geol. Soc. Japan,* 74 (11), 589-594. (J+E)
_____ (1971a), Classification of sandstone: analysis and proposal. *J. Geol.,* **79** (5), 509-525.
_____ (1971b), Again on classification and nomenclature of sandstones. *J. Geol. Soc. Japan,* 77 (6), 395-396. (J)
_____ (1971c), Geosynclinal clastic sediments and their depositional environments. *Special Issue, Soc. Mining Geol. Japan,* (4), 81-96. (J+E)
_____ (1973), Abyssal feldspathic sediments in the northwestern Pacific. *Init. Repts. DSDP,* **20**, 359-362.
_____ (1974), Migration of ancient arc-trench systems. *Soc. Econ. Paleont. Mineral., Spec. Publ.,* (19), 311-320.
_____ (1977), Preliminary study of sandstones of the Shimanto Supergroup in Kyushu, with special reference to "Petrographic Zone". *Sci. Rep. Dept. Geol. Kyushu Univ.,* **12** (3), 203-214. (J+E)
_____ (1983), Collision orogenesis and sedimentation in Hokkaido, Japan. *In*: Hashimoto, M. and Uyeda, S. (eds.), *Accretion Tectonics in the Circum-Pacific Regions.* Terra Scientific Publishing Co., Tokyo, 91-105.
_____ (1988), The face of the trench. *Geology Today,* **4**, 167-171.
_____ (1992), For the establishment of sequence stratigraphy in Japan. *J. Sed. Soc.Japan,* (36), 5-8. (J+E)
_____ (1997), Cretaceous petroprovinces in the Japanese Islands and their tectonic significance. *J. Geol. Soc. Philippines,* **52** (3-4), 285-297.
_____ (1998a), A short history of sedimentology – International aspect. *J. Sed. Soc. Japan,* (47), 3-15. (J+E)
_____ (1998b), A short history of sedimentology in Japan. *J. Sed. Soc. Japan,*, (48), 5-12. (J+E)
_____ (2000), Nature and development of Cretaceous sedimentary basins in East Asia: a review. *Geosci. J.,* **4** (4), 271-282.
_____ (2002), *Sedimentology. A Way to the New Discipline of Earth Sciences.* Kokon Shoin Publishers, Tokyo. (J)
_____ (2003), Sedimentology: History in Japan. *In*: Middleton, G. V. (ed.), *Encyclopedia of Sediments and Sedimentary Rocks.* Kluwer Academic Pub., Dordrecht, 635-637.
_____ (2004), Founders of sedimentology (6). Henry Clifton Sorby – Father of Sedimentology. *J. Sed. Soc. Japan,* (58), 115-120. (J)
_____ and Kobayashi, K. (1974), Palaeoenvironments of the ocean floor based on sedimentological evidence. *Kagaku (Science),* **44**(4), 212-221. (J)
_____ and Mateer, N. J. (eds.) (2000), *Cretaceous Environments of Asia.* Elsevier, Amsterdam.
_____ and Nakao, S. (1968), Plagioclase arenite in Lower Cretaceous flysch in the Furano area, Hokkaido. *J. Geol. Soc. Japan,* 74, 451-152. (J)
_____ and Ohta, S. (1993), Photographic evidence of variable bottom-current activity in the Suruga and Sagami Bays, central Japan. *Sed. Geol.,* **82**, 221-237.
_____ and Sakai, T. (2000), The Cretaceous System of the Japanese Islands and its physical environments. *In*: Okada, H. and Mateer, N. J. (eds.) (2000), *Cretaceous Environments of Asia.* Elsevier, Amsterdam, 113-144.
_____ and Tandon, S. K. (1984), Resedimented conglomerates in a Miocene collision suture,

Hokkaido, Japan. *In*: Koster, E. H. and Steel, R. J. (eds.), *Sedimentology of Gravels and Conglomerates. Canad. Soc. Petrol. Geol. Bull.*, **10**, 413-427.

Okami, K. (1973), Sedimentological study of the Iwaki Formation of the Joban Coal Field. *Sci. Repts. Tohoku Univ., 2nd Ser.*, **44** (1), 1-53.

Okazaki, H. (1999), Backset bedding exampled from the Pleistocene Kamiizumi Formation. *J. Sed. Soc. Japan*, (49), 1-5. (J)

_____ and Masuda, F. (1992), Depositional systems of the Late Pleistocene sediments in Paleo-Tokyo Bay area. *J. Geol. Soc. Japan*, **98** (3), 235-258. (J+E)

Omori, M. (1968), Existing circumstances of the long-term geological research projects. *In*: Geological Society of Japan, *Geological Sciences in Japan – Past, Present and Future. Jubilee Publication in Commemoration of the Seventy-fifth Anniversary of the Geological Society of Japan*. The Geological Society of Japan, 592-603. (J)

_____ (1993), *Trace Fossils (by R.G. Bromley)*. Tokai Univ. Press, Tokyo. (J)

Open University (1978), *Oceanography. The Changing Oceans*. Open Univ. Press, Milton Keynes.

Oppel, A. (1856-1858), *Die Juraformation Englands, Frankreichs und des süd-Westlichen Deutschlands*. Stuttgart.

Ota, M. (1968), The Akiyoshi Limestone Group: A geosynclinal organic reef complex. *Bull. Akiyoshi-dai Sci. Mus.*, (5), 1-44. (J+E)

Ozawa, Y. (1923), Stratigraphical studies of the so-called Chichibu Palaeozoic including the Akiyoshi limestone. *J. Geol. Soc. Japan*, **30** (357), 227-243. (J+E)

Packham, G. H. (1954), Sedimentary structures as an important factor in the classification of sandstones. *Am. J. Sci.*, **252**, 466-476.

Pantin, H. M. and Leeder, M. R. (1987), Reverse flow in turbidity currents: the role of internal solitons. *Sedimentology*, **34**, 1143-1155.

Paola, C., Mullin, J., Ellis, C., Mohrig, D. C., Swenson, J. R., Parker, G., Hickson, T., Heller, P. L., Pratson, L., Wyvitski, J., Sheets, B. and Strong, N. (2001), Experimental stratigraphy. *GSA Today*, **11** (7), 4-9.

Parea, G. C. and Ricci-Lucchi, F. (1975), Turbidite key-beds as indicators of ancient deep sea plains. *Proc. IX Int. Congr. Sed., Nice, Theme 1*, **1**, 235-245.

Park, C. (2001), *The Environment. Principles and Applications. 2nd ed.* Routledge, New York.

Payton, C. E. (ed.)(1977), Seismic stratigraphy – Application to hydrocarbon exploration. *Am. Assoc. Petrol. Geol. Mem.*, **26**.

Pelletier, B. R. (1958), Pocono paleocurrents in Pennsylvania and Maryland. *Geol. Soc. Am. Bull.*, **69**, 1033-1064.

Pettijohn, F. J. (1941), Persistence of heavy minerals and geologic age. *J. Geol.*, **49**, 610-625.

_____ (1949), *Sedimentary Rocks*. Harper & Sons, New York.

_____ (1954), Classification of sandstones. *J. Geol.*, **62**, 360-365.

_____ (1957a), *Sedimentary Rocks. 2nd Ed.* Harper & Sons, New York.

_____ (1957b), Paleocurrents of Lake Superior Precambrian quartzites. *Geol. Soc. Am. Bull.*, **68**, 409-480.

_____ (1962), Paleocurrents and paleogeography. *Am. Assoc. Petrol. Geol. Bull.*, **46**, 1468-1493.

_____ (1975), *Sedimentary Rocks. 3rd Ed.* Harper & Sons, New York.

_____ (1984), *Memoirs of an Unrepentant Field Geologist. A Candid Profile of Some Geologists and Their Science, 1921-1981*. Univ. Chicago Press, Chicago.

_____ Potter, P. E. and Siever, R. (1972), *Sand and Sandstone*. Springer-Verlag, New York.

Pickering, K. T. and Hiscott, R. N. (1985), Contained (reflected) turbidity currents from the Middle Ordovician Cloridorme Formation, Quebec, Canada: an alternative to the antidune hypothesis. *Sedimentology*, **32**, 373-394.

Playfair, J. (1802), *Illustrations of the Huttonian Theory of the Earth*. Cadell and Davies, London.

Plint, A. G., Eyles, N., Eyles, C. H. and Walker, R. G. (1992), Control of sea level change. *In*:

Walker, R. G. and James, N. P. (eds.), *Facies Models: Response to Sea Level Change.* Geol. Assoc. Canada, St. Johns, 15-26.

Plumley, W.J. (1948), Black Hills Terrace gravels: A study in sediment transport. *J. Geol.*, **56**, 526-577.

Porter, J. J. (1962), Electron microscopy of sand surface texture. *J. Sed. Petrol.*, **32**, 124-135.

Posamentier, H. W., Allen, G. P. and James, D. P. (1990), Aspects of sequence stratigraphy: recent and ancient examples of forced regression (abs.). *Am. Assoc. Petrol. Geol. Bull.*, **74**, 742.

_____ _____ _____ and Tesson, M. (1992), Forced regressions in a sequence stratigraphic framework: concepts, examples, and exploration significance. *Am. Assoc. Petrol. Geol. Bull.*, **76** (11), 1687-1709.

Potter, P. E. (1962), Regional distribution pattern of Pennsylvanian sandstones in Illinois basin. *Bull. Am. Assoc. Petrol. Geol.*, **46**, 1890-1911.

_____ (1967), Sand bodies and sedimentary environments. A review. *Bull. Am. Assoc. Petrol. Geol.*, **51**, 337-365.

_____ (1978), Sedimentology – yesterday, today, and tomorrow. *In:* Fairbridge, R. W. and Bourgeois, J. (eds.): *The Encyclopedia of Sedimentology.* Dowden, Hutchinson & Ross, Stroudsburg, Pa., 718-724.

_____ Maynard, J. B. and Pryor, W. A. (1980), *Sedimentology of Shale*. Springer-Verlag, New York.

_____ and Olson, J. S. (1954), Variance components of cross-bedding direction in some basal Pennsylvanian sandstones of the Eastern Interior Basin: Geological application. *J. Geol.*, **62**, 50-73.

_____ and Pettijohn, F. J. (1963), *Paleocurrents and Basin Analysis.* Springer Verlag, Berlin.

Powers, M. C. (1953), A new roundness scale for sedimentary particles. *J. Sed. Petrol.*, **23**, 117-119.

Pryor, W. A. (1971), Grain shape. *In*: Carver, R. E. (ed.), *Procedures in Sedimentary Petrology.* Wiley-Interscience, New York, 131-150.

Pustovalov, L. V. (1933), Geokhimickeskie fatsii I ikh znachenie v obshchei I prikladnoi geologii. *Problemy Sovietskoi Geologii*, **1**, 57-80.

Reading, H.G. (1972), Global tectonics and the genesis of flysch successions. *24th IGC, Sec. 6*, 59-66.

_____ (1974), What have we learned in sedimentology in the past seventy years? *Geol. Mag.*, **111** (5), 455-461.

_____ (1978), *Sedimentary Environments and Facies.* Elsevier, New York.

_____ (1986), *Sedimentary Environments and Facies. 2nd Ed.* Blackwell, Oxford.

Reiche, P. (1943), Graphic representation of chemical weathering. *J. Sed. Petrol.*, **13** (2), 58-68.

Reineck, H.-E. and Singh, I. B. (1973), *Depositional Sedimentary Environments*. Springer-Verlag, Berlin.

Research Group on Clastic Sediments (ed.) (1983), *Study Methods of Sediments – Conglomerates, Sandstones and Mudstones*. Association for Geological Collaboration. (J)

Ricci-Lucchi, F. (1995), *Sedimentographica: Photographic Atlas of Sedimentary Structures. 2nd Ed.* Columbia Univ. Press, New York.

Rich, J. L. (1923), Shoestring sands of eastern Kansas. *Bull. Am. Assoc. Petrol. Geol.*, 7, 103-113.

_____ (1951), Three critical environments of deposition and criteria for recognition of rocks deposited in each of them. *Geol. Soc. Am. Bull.*, **62**, 1-20.

Richardson, M. J., Wimbush, M. and Mayer, L. (1981), Exceptionally strong near-bottom flows on the continental rise of Nova Scotia. *Science*, **213**, 887-888.

Rifardi, Oki, K. and Tomiyasu, T. (1998), Sedimentary environments based on textures of surface sediments and sedimentation rates in the South Yatsushiro Kai (Sea), southwest Kyushu, Japan. *J. Sed. Soc. Japan*, (48), 67-84.

Rittenhouse, G. (1943), Transportation and deposition of heavy minerals. *Geol. Soc. Am. Bull.*,

**54**, 1725-1780.
_____ (1971), Pore-space reduction, solution and cementation. *Am. Assoc. Petrol. Geol. Bull.*, **55**, 80-91.
Rivière, A. (1952), Expression analytique générale de la granulométrie des sédiments meubles. *Bull. Soc. Géol. France, Sér. 6*, **2**, 155-167.
Rona, P. A. (1973), Worldwide unconformities in marine sediments related to eustatic changes of sea level. *Nature Phys. Sci.*, **244**, 25-26.
Roser, B. P. and Korsch, R. J. (1986), Determination of tectonic setting of sandstone-mudstone suites using $SiO_2$ content and $K_2O/Na_2O$ ratio. *J. Geol.*, **94**, 635-650.
Rubey, W. W. (1933), Settling velocities of gravel, sand and silt particles. *Am. J. Sci.*, **224**, 325-338.
Rusnak, G. A. (1957), The orientation of sand grains under conditions of unidirectional fluid flow. *J. Geol.*, **65**, 384-409.
Russell, R. J. (1965), Memorial to Harold Norman Fisk. *Geol. Soc. Am., Proc.* (1965), 53-58.
Saito, Ya. (1977), Petrogenesis of bedded chert of the Triassic Adoyama Formation. *Bull. Natn. Sci. Mus. Tokyo, Ser. C*, **3** (3), 151-156.
_____ and Imoto, N. (1978), Chertification of siliceous sponge spicule deposit. *Mem. Geol. Soc. Japan*, (15), 91-102. (J+E)
Saito, Yo. (1989), Classification of shelf sediments and their sedimentary facies in the storm-dominated shelf: a review. *J. Geogr.*, **98** (3), 350-365. (J+E)
_____ (1991), Sequence stratigraphy on the shelf and upper slope in response to the latest Pleistocene-Holocene sea-level changes off Sendai, northeast Japan. *Int. Assoc. Sedimentol. Spec. Publ.*, **12**, 133-150.
_____ (1994), Shelf sequence and characteristic bounding surfaces in a wave-dominated setting: Latest Pleistocene-Holocene examples from Northeast Japan. *Marine Geol.*, **120**, 105-127.
_____ Hoyanagi, K. and Ito, M. (1995), Sequence stratigraphy: Toward a new dynamic stratigraphy. *Mem. Geol. Soc. Japan*, (45), 249. (J+E, 2 papers in E)
Sakai, A. (1992), Local variation of sandstone compositions in the southern belt of the Chichibu Terrane. *Mem. Geol. Soc. Japan*, (38), 227-236. (J+E)
Sakai, H. (1988a), Toi-misaki olistostrome of the Southern Belt of the Shimanto Terrane, South Kyushu. I. Reconstruction of depositional environments and stratigraphy before the collapse. *J. Geol. Soc. Japan*, **94** (10), 733-747. (J+E)
_____ (1988b), Toi-misaki olistostrome of the Southern Belt of the Shimanto Terrane, South Kyushu. II. Deformation structures of huge submarine slides and their processes of formation. *J. Geol. Soc. Japan*, **94** (11), 837-853. (J+E)
_____ (1988c), Origin of the Misaki Olistostrome Belt and re-examination of the Takachiho Orogeny. *J. Geol. Soc. Japan*, **94** (12), 945-961. (J+E)
Sakai, T. (1978), Geologic structure and stratigraphy of the Shimantogawa Group in the middle reaches of the Gokase River, Miyazaki Prefecture. *Sci. Repts., Dept. Geol., Kyushu Univ.*, **13** (1), 23-38. (J+E)
_____ Kusaba, T., Nishi, H., Komori, M. and Watanabe, M. (1987), Olistostrome of the Shimanto Terrane in the Nichinan area, southern part of the Miyazaki Prefecture, South Kyushu – with reference to deformation and mechanism of emplacement of olistoliths. *Sci. Repts., Dept. Geol., Kyushu Univ.*, **15** (1), 167-199. (J+E)
_____ and Okada, H. (1997), Sedimentation and tectonics of the Cretaceous sedimentary basins of the Axial and Kurosegawa Tectonic Zones in Kyushu, SW Japan. *Mem. Geol. Soc. Japan*, (48), 7-28.
Sakakura, K. (1964), *Coal Geology*. Gijutsu Shoin, Tokyo. (J)
Sakamoto, T. (1950), The origin of the Pre-Cambrian banded iron ores. *Am. J. Sci.* **248**, 449-474.

Sander, B. (1912), Über tektonische Gesteinsfazies. K. K. Geol. Reichsanstalt. Verhand., 249-257.
Sanders, J. E. (1963), Concepts of fluid mechanics provided by primary sedimentary structures. *J. Sed. Petrol.*, **33**, 173-179.
Sano, H. (1988a), Permian oceanic rocks of Mino Terrane, central Japan. Part I. Chert facies. *J. Geol. Soc. Japan*, **94** (9), 697-709.
_____ (1988b), Permian oceanic rocks of Mino Terrane, central Japan. Part II. Limestone facies. *J. Geol. Soc. Japan*, **94** (12), 963-976.
_____ (1989), Permian oceanic rocks of Mino Terrane, central Japan. Part III. Limestone breccia facies. *J. Geol. Soc. Japan*, **95** (7), 527-540.
Sato, D. (1924), *Petrographic Geology*. Ogiwara Seibunkan, Tokyo. (J)
_____ (1940), *Revised Petrographic Geology*. Ogiwara Seibunkan, Tokyo. (J)
Sato, Y. (1969), Geological significance of zircon-garnet-tourmaline ratio of the Paleogene sandstones of northwestern Kyushu, Japan. *Geol Surv. Japan Rept.*, (235), 1-46.
Schlanger, S. O. and Jenkyns, H. C. (1976), Cretaceous oceanic anoxic events: causes and consequences. *Geol. Mijnbouw*, **55**, 179-184.
Schmitz, J. W., Jr. (1984), Abyssal eddy kinetic energy in the North Atlantic. *J. Mar. Res.*, **42**, 509-536.
Sclater, J. G., Anderson, R. N. and Bell, M. L. (1971), Elevation of ridges and evolution of the central eastern Pacific. *J. Geophys. Res.*, **76**, 7888-7915.
Sedimentological Society of Japan (ed.) (1998), *Encyclopaedia of Sedimentology*. Asakura Shoten, Tokyo. (J)
Seely, D. R., Vail, P. R. and Walton, G. G. (1974), Trench slope model. *In*: Burk, C. A. and Drake, C. L. (eds.), *The Geology of Continental Margin*. Springer-Verlag, Berlin, 249-260.
Seibold, E. and Seibold, I. (2002), Sedimentology: from single grains to recent and past environments: some trends in sedimentology in the twentieth century. *In*: Oldroyd, D. R. (ed.), *The Earth Inside and Out: Some Major Contributions to Geology in the Twentieth Century*. Geological Society, London, Special Publications, **192**, 241-250.
Seilacher, A. (1967), Bathymetry of trace fossils. *Marine Geol.*, **5** (5-6), 413-428.
_____ (1978), Use of trace fossils for recognizing depositional environments. *In*: Basan, P. B. (ed.), *Trace Fossils Concepts*. Soc. Econ. Paleont. Mineral., Short Course, (5), 175-201.
Selley, R. C. (1976), *An Introduction to Sedimentology*. Academic Press, London.
Sengupta, S.(1994), *Introduction to Sedimentology*. A. A. Balkema, Rotterdam.
Seuss, E. (1875), *Die Entstehung der Alpen*. W. Braunmüller, Wien.
Shanmugam, G. (2000), 50 years of the turbidite paradigm (1950s-1990s). *Marine and Petroleum Geol.*, **17**, 285-342.
_____ (2002), Ten turbidite myths. *Earth-Science Rev.*, **58** (3-4), 311-341.
Sharp, R. P. (1978), Martian sedimentation. *In*: Fairbridge, R.W. and Bourgeois, J. (eds.), *The Encyclopedia of Sedimentology*. Dowden, Hutchinson & Ross, Inc., Stroudsburg, Pa., 475-479.
Sheldon, R. W., Evelyn, T. P. T. and Parsons, T. R. (1967), On the occurrence and formation of small particles in seawater. *Limnol. Oceanogr.*, **12** (3), 367-375.
Shepard, F. P. (1967), *Submarine Geology. 2nd Ed.* Harper, New York.
_____ and Young, R. (1961), Distinguishing between beach and dune sands. *J. Sed. Petrol.*, **31**, 196-214.
Shiki, T. (1959), Studies on sandstones in the Maizuru Zone, Southwest Japan. I. Importance of some relations between mineral compositions and grain size. *Mem. Coll. Sci., Univ. Kyoto, Ser. B*, **25**, 239-246.
_____ (1993), Study of event sediments. *Earth Monthly, Spec. Issue*, (8), 227-244. (J)
_____ and Mizutani, S. (1965), On "Graywacke" – Part I. Review of studies on the characteristics and the definitions of graywacke – . *Chikyu-kagaku (Earth Science)*, (81), 21-32. (J+E)

_____ and Yamazaki, T. (1996), Tsunami-induced conglomerates in Miocene upper bathyal deposits, Chita Peninsula, central Japan. *Sediment. Geol.*, **104** (1-4), 175-188.
_____ Chough, S. and Einsele, G. (eds.) (1996), Marine sedimentary events and their records. *Sediment. Geol.*, **104** (1-4), 1-255.
Shimazaki, H., Shindo, S. and Yoshida, S. (eds.) (1995), *Geological Basis for Radioactive Waste Disposal.* Univ. Tokyo Press, 389p. (J)
Shimizu, D. (1996), *History of Earth Sciences described in Classics.* Tokai Univ. Press, Tokyo. (J)
Shoji, R. (1955), A fundamental study on sedimentation – Cyclic sedimentation and ripple mark in a settling pond – . *J. Geol. Soc. Japan*, **61** (722), 518-531. (J+E)
_____ (1960), On the origin of the cyclothemic arrangement in the coal bearing strata of Japan (1)(2). *J. Geol. Soc. Japan*, **66** (781), 660-674; (782), 733-741. (J+E)
_____ (1966), Recent progress in sedimentology. *J. Geogr.*, 75 (1/2), 12-35, 70-83. (J)
_____ (1971a), *Sedimentology.* Asakura Shoten, Tokyo. (J)
_____ (1971b), *Sedimentary Petrology.* Asakura Shoten, Tokyo. (J)
_____ Taguchi, K. and Iijima, A. (1968), Sedimentology in Japan, a review. *In*: Geological Society of Japan, *Geological Sciences in Japan – Past, Present and Future. Jubilee Publication in Commemoration of the Seventy-fifth Anniversary of the Geological Society of Japan.* The Geological Society of Japan, 99-122. (J)
Shrock, R. R. (1948), *Sequence in Layered Rocks.* McGraw Hill, New York.
_____ (1957), William Henry Twenhofel, 1875-1957. *Am. Assoc. Petrol. Geol. Bull.*, **41**, 978-980.
Simons, D. B., Richardson, E. V. and Nordin, C. F. (1965), Sedimentary structures generated by flow in alluvial channels. *Soc. Econ. Paleont. Mineral., Spec. Publs.*, **12**, 34-52.
Simons, E. N. (1952), Henry Clifton Sorby, a pioneer in metallurgical microscopy. *Metallurgia*, **46**, 199-200.
Simpson, J. E. (1987), *Gravity Currents in the Environment and the Laboratory. Ellis Horwood Series in Environmental Science.* John Wiley & Sons, Ltd., Chichester.
Sliter, W. V. (1977), Cretaceous foraminifers from the southwestern Atlantic Ocean, Leg 36, Deep Sea Drilling Project. *Init. Rep. DSDP*, **36**, 519-573.
Sloss, L. L. (1963), Sequences in the cratonic interior of North America. *Geol. Soc. Am. Bull.*, 74, 93-114.
_____ Krumbein, W. C. and Dapples, E. C. (1949), Integrated facies analysis. *In*: Longwell, C. R. (ed.), *Sedimentary Facies in Geologic History.* Geol. Soc. Am. Mem., **39**, 9-123.
Smith, W. (1815), *A Map of the Strata of England and Wales, with Part of Scotland.* J. Cary, London.
_____ (1816), *Strata Identified by Organised Fossils. Containing Prints on Coloured Paper of the Most Characteristic Specimens in Each Stratum.* W. Arding, London.
Smithson, F. (1941), The alteration of detrital minerals in the Mesozoic rocks of Yorkshire. *Geol. Mag.*, **78**, 97-112.
Sneed, E. D. and Folk, R. L. (1958), Pebbles in the lower Colorado River, Texas, a study in particle morphogenesis. *J. Geol.*, **66**, 114-150.
Soh, W. (1985), Sedimentary facies and processes of paleo-submarine channels in the Mio-Pliocene Minobu Formation, Fujikawa Group, central Japan. *J. Geol. Soc. Japan*, **91**, 87-107. (J+E)
Sonnenfeld, P. (1985), Models of Upper Miocene evaporite genesis in the Mediterranean region. *In*: Stanley, D. J. and Wezel, F.-C. (eds.), *Geological Evolution of the Mediterranean Basin.* Springer-Verlag, New York, 323-346.
_____ and Finetti, I. (1985), Messinian evaporites in the Mediterranean: a model of continuous inflow and outflow. *In*: Stanley, D. J. and Wezel, F.-C. (eds.), *Geological Evolution of the Mediterranean Basin.* Springer Verlag, New York, 347-353.
Sorby, H. C. (1849), On the geology of the Malvern Hills. *Sheffield Lit. and Phil. Soc. Rep.*, 19.

_____ (1851), On the microscopical structure of the calcareous grit of the Yorkshire coast. *Quart. J. Geol. Soc. London*, **8**, 1-6.
_____ (1852), On the oscillation of the currents drifting the sandstone beds of the south-east of Northumberland, and on their general direction in the coalfield in the neighbourhood of Edinburgh. *Proc. West Yorks. Geol.Soc.*, **3**, 232-240.
_____ (1859), On the structures produced by the currents present during the deposition of stratified rocks. *Geologist*, **2**, 137-147.
_____ (1877), The application of the microscope to geology. *Mon. Microscop. J.*, **17**, 113-136.
_____ (1879), The structure and origin of limestones. *Quart. J. Geol. Soc. London*, **35**, 56-95.
_____ (1880), The structure and origin of non-calcareous stratified rocks. *Quart. J. Geol. Soc. London*, **36**, 46-92.
_____ (1908), On the application of quantitative methods to the study of the structure and history of rocks. *Quart. J. Geol. Soc. London*, **64**, 171-233.
Southard, J. B., Young, R. A. and Hollister, D. C. (1971), Experimental erosion of calcareous ooze. *J. Geophys. Res.*, **76**, 5903-5909.
Spudis, P. D. (1996), *The Once and Future Moon*. Smithonian Inst. Press, Washington, D.C.
_____ (2004), The New Moon. *Scientific American*, **289** (6), 86-93.
Stanley, D. J. (1993), Model for turbidite-to-contourite continuum and multiple process transport in deep marine settings: examples in the rock record. *Sed. Geol.*, **82**, 241-255.
Steno, N. (1669), *De solido intra solidum naturaliter contento dissertationis prodromus*. Florence.
Sternberg, H. (1875), Untersuchungen über Längenund Querprofil geschiebeführender Flüsse. *Zeitschr. Bauwesen*, **25**, 483-506.
Stille, Hans (1936), Wege und Ergebnisse der geologisch-tektonischen Forschung. *Festchr. Kaiser-Wilhelm Gesell. Förd. Wiss.*, **2**, 77-97.
Stokes, G. G. (1851), On the effect of the internal friction of fluids on the motion of pendulums. *Trans. Cambridge Philos. Soc.*, **9** (2), 8-106.
Stommel, H. (1958), The abyssal circulation. *Deep-Sea Res.*, **5**, 80-82.
_____ and Aarons, A.B. (1960), On the abyssal circulation of the world ocean II. *Deep-Sea Res.*, **6**, 217-233.
Stow, D. A. W. (1985), Deep-sea clastics: where are we and where are we going? *In*: Brenchley, P. J. and Williams, B. P. J. (eds.), *Sedimentology. Recent Developments and Applied Aspects*. The Geol. Soc. London, 67-93.
_____ and Lovell, J. P. B. (1979), Contourites: Their recognition in modern and ancient sediments. *Earth-Sci. Review*, **14**, 251-291.
Strahler, A. N. (1958), Dimensional analysis applied to fluvially eroded landforms. *Geol. Soc. Am. Bull.*, **69**, 279-300.
Strakhov, N. M. (1962), *Osnovi teorii litogeneza, tom 3, Zakonomernosti sostava i razmesicheniya aridnikh otladzenii [Principles of Lithogenesis, v. 3, Compositional and Distributional Regularities of Arid Sediments]*. Akademii Nauk SSSR, Geol. Inst., Moscow.
_____ (1970), Evolution of concepts of lithogenesis in Russian geology (1880-1970). *Lith. Mineral Res.*, (2), 157-177.
Stubblefield, C. J. (1965), Edward Battersby Bailey 1881-1965. *Biographical Memoirs of Fellows of the Royal Society*, **11**, 1-21.
Suess, E. (1875), *Die Entstehung der Alpen*. W. Braumüller, Wien.
Sullivan, W. and Bell, J. (2004), The MarsDial: A Sundial for the Red Planet. *Planetary Report,* **24** (1), 6-11.
Summerson, C. H. (ed.) (1976), *Sorby on Sedimentology. A Collection of Papers from 1851 to 1908 by Henry Clifton Sorby*. Comparative Sedimentology Laboratory, University of Miami.
Sun, G., Zheng, S. L., Dilcher, D. L., Wang, Y. D. and Mei, W. W. (2001), *Early Angiosperms and Their Associated Plants from Western Liaoning, China*. Shanghai Sci. Tech. Educ. Publ. House.

Sundborg, Å. (1956), The River Klarälven: a study of fluvial processes. *Geogr. Ann.*, **38**, 127-316.
Suzuki, K. (1994), Three dimensional morphology and sedimentary structures of the floodplain deposits of the Yasu River, central Japan, at the flood of September 9, 1993 – A few inversely graded layer units formed by a single flood. *J. Geol. Soc. Japan*, **100** (11), 867-875. (J+E)
Suzuki, N. and Taguchi, K. (1982), Distribution and diagenesis of fatty acids in ancient marine sediments of Miocene age from the Shinjo Oil Field, northeastern Japan. *J. Geol. Soc. Japan*, **88** (3), 185-198. (J+E)
Suzuki, S., Asiedu, D. K. and Shibata, T. (1997), Composition of sandstones of the Kenseki Formation and paleogeographic reconstruction in the Lower Cretaceous, Innerside of Southwest Japan. *J. Geol. Soc. Philippines*, **52**, 143-159.
Suzuki, Y. (2003), Historical review of the study on intermediate and deep earthquakes. *JAHIGEO Bull.*, (21), 15-24. (J)
Swift, D. J. P. (1968), Coastal erosion and transgressive stratigraphy. *J. Geol.*, **76**, 444-456.
Tada, R. (1994), Paleoceanographic evolution of the Japan Sea. *Palaeogeography Palaeoclimatology Palaeoecology*, **108**, 487-508.
Taguchi, K. (ed.) (1983), Diagenesis in sediments. *J. Sed. Soc. Japan*, (17/18/19), 1-196. (J+E)
Taira, A. (1981), Process of the formation of the Shimanto Belt. *Kagaku (Science)*, **51**, 516-523. (J)
_____ and Tashiro, M. (eds.) (1980), *Geology and Palaeontology of the Shimanto Belt.* Rinyakosaikai Press, Kochi. (J+E)
_____ Okada, H., Whitaker, J. H. McD. and Smith, A. J. (1982), The Shimanto Belt of Japan: Cretaceous-lower Miocene active-margin sedimentation. *In*: Leggett, J. K. (ed.), *Trench-forearc Geology. Geol. Soc., Spec. Publ.*, **10**, 5-26.
Taira, K. and Teramoto, T. (1985), Bottom currents in Nankai Trough and Sagami Trough. *J. Oceanogr. Soc. Japan*, **41**, 388-398.
Takahashi, J. (1922), The marine kerogen shales from the oil fields of Japan. A contribution to the origin of petroleum. *Sci. Repts. Tohoku Imp. Univ., Ser. III*, **1** (2), 63-156.
_____ (1939), Synopsis of glauconitization. *In*: Trask, P. D. (ed.), *Recent Marine Sediments.* Am. Assoc. Petrol. Geol., Tulsa, Okla., 503-512.
_____ (1940), Sedimentary cycles in the Oga Series (I) (II). *J. Japan. Assoc. Min. Petr. Econ. Geol.*, **23** (2), 70-78; (5), 194-203. (J)
_____ and Yagi, T. (1929a), The peculiar mud-grains in the recent littoral and estuarine deposits, with special reference of the origin of glauconite. *Econ. Geol.*, **24**, 838-852.
_____ _____ (1929b), Characteristic particles in coastal marine sediments in Japan. *J. Japan. Assoc. Min. Petr. Econ. Geol.*, **1** (2), 64-71. (J)
Takano, K. and Hara, H. (1970), A preliminary analysis of current meter records. *La Mer*, **8**, 205-228.
Takano, O. (1990), Depositional processes of trough-fill turbidites of the upper Miocene to Pliocene Tamugigawa Formation, northern Fossa Magna, central Japan. *J. Geol. Soc. Japan*, **96**, 1-17. (J+E)
_____ (2000), Sequence stratigraphy as a genetic tool for analyzing sedimentary basins: methods and case-studies. *Fukada Geol. Inst., Fukadaken Library*, (36), 1-61. (J)
_____ (2001), Recognition of depositional sequences in terrestrial strata – a review of developments in terrestrial sequence stratigraphy. *J. Japan. Assoc. Petrol. Technol.*, **66** (3), 332-341. (J+E)
Takayanagi, Y. (1999), *Challenger at Sea. A Ship that revolutionized Earth Science by Kenneth J. Hsü.* Tokai Univ. Press. (J+E)
_____ Saito, T., Okada, H., Ishizaki, K., Oda, M., Hasegawa, S., Okada, Hisatake and Manickam, S. (1985), Late Quaternary palaeoenvironment of the coastal water, east of Honshu, based on analysis of material from offshore drilling. *In*: Kajiura, K. (ed.), *Ocean*

*Characteristics and their Changes*. Koseisha-Koseikaku, Tokyo,, 397-412.

Takeuchi, M. (1986), Detrital garnet in Paleozoic–Mesozoic sandstone from the central part of the Kii Peninsula. *J. Geol. Soc. Japan*, **92**, 289-306. (J+E)

_____ (2000), Message from chemistry of clastic grains: Provenance analysis based on chemical composition of detrital garnet and tectonic events in East Asia. *Mem. Geol. Soc. Japan*, (57), 183-194.

Takizawa, F. (1985), Jurassic sedimentation in the South Kitakami Belt, Northeast Japan. *Bull. Geol. Surv. Japan*, **36** (5), 203-320.

Tanaka, J. (1989), Sedimentary facies of the Cretaceous Izumi turbidite system, Southwest Japan – an example of turbidite sedimentation in an elongated strike-slip tectonic basin. *J. Geol. Soc. Japan*, **95** (2), 119-128.

Tanaka, K. (1963), A study on the Cretaceous sedimentation in Hokkaido, Japan. *Rep. Geol. Surv. Japan*, (197), 1-119.

_____ (1965), Izumi Group in the central part of the Izumi Mountain range, Southwest Japan, with special reference to the sedimentary facies and cyclic sedimentation. *Rep. Geol. Surv. Japan*, (212), 1-34. (J+E)

_____ (1970), Sedimentation of the Cretaceous flysch sequence in the Ikushum-betsu area, Hokkaido, Japan. *Rep. Geol. Surv. Japan*, (236), 1-102.

Tanaka, K. L., Banerdt, W. B., Kargel, J. S. and Hoffman, N. (2001), Huge, $CO_2$-charged debris-flow deposit and tectonic sagging in the northern plains of Mars. *Geology*, **29** (5), 427-430.

Tanton, T. L. (1930), Determination of age-relations in folded rocks. *Geol. Mag.*, **67**, 73-76.

Tarduno, J. A., Sliter, W. V., Kroenke, L., Leckie, M., Mayer, H., Mahoney, J. J., Musgorave, R., Storey, M. and Minterer, E. L. (1991), Rapid formation of the Ontong Java Plateau by Aptian mantle plume volcanism. *Science*, **254**, 399-403.

Tateishi, M. (1978), Sedimentology and basin analysis of the Paleogene Muro Group in the Kii Peninsula, Southwest Japan. *Mem. Fac. Sci., Kyoto Univ., Ser. Geol. Mineral.*, **14** (2), 187-232.

Tayama, R. (1950), Sumarine topography and distribution of bottom sediments on eastern part of Sagami Bay. *Hydrographic Bull., Maritime Safety Agency, Tokyo*, **17**, 1-17. (J+E)

Teichert, C. (1958), Concept of facies. *Bull. Am. Assoc. Petrol. Geol.*, **42**, 2718-2744.

Teraoka, Y. (1970), Cretaceous formations in the Onogawa basin and its vicinity, Kyushu, Southwest Japan. *Rep. Geol. Surv. Japan*, (237), 1-84. (J+E)

_____ (1979), Provenance of the Shimanto geosynclinal sediments inferred from sandstone compositions. *J. Geol. Soc. Japan*, **85** (12), 753-769. (J+E)

Thiel, G. A. (1940), The relative resistance to abrasion of mineral grains of sand. *J. Sed. Petrol.*, **10**, 103-124.

Thomas, H. H. (1902), The mineralogical constitution of the finer material of the Bunter pebble beds in the west of England. *Quart. J. Geol. Soc. London*, **58**, 620-632.

Tickell, R. G. (1924), Correlative value of the heavy minerals. *Am. Assoc. Petrol. Geol. Bull.*, **8**, 158-168.

Tokuhashi, S. (1979), Three dimensional analysis of a large sandy-flysch body, Mio-Pliocene Kiyosumi Formation, Boso Peninsula, Japan. *Mem. Fac. Sci., Kyoto Univ., Ser. Geol. Mineral.*, **46** (1), 1-60.

Tokuoka, T. (1970), Orthoquartzitic gravels in the Paleogene Muro Group, Southwest Japan. *Mem. Fac. Sci., Kyoto Univ., Ser. Geol. Mineral.*, **37** (1), 113-132.

_____ (1993), Environmental geology of brackish water lakes, with special reference to Nakaumi and Shinji Lakes. *In*: Committee of Environmental Geology, Geological Society of Japan (ed.), *Some Problems of Geo-Environments based on Environmental Geology*. Tokai Univ. Press, Tokyo, 61-88. (J)

_____ (1999), Natural history of brackish-water and marsh areas and their development and conservation. *Fukada Geol Inst., Fukadaken Library*, (30), 1-72. (J)

_____ Anma,K. and Inouchi, Y. (eds.) (1993), Lakes – Origin, environment and geology II. *Mem. Geol. Soc. Japan*, (39), 1-189. (J+E)

Tomiyasu, T., Nagano, A., Yonehara, N., Sakamoto, H., Rifardi, Oki, K. and Akagi, H.. (2000), Mercury contamination in the Yatsushiro Sea, south-western Japan: spatial variations of mercury in sediment. *Sci. Total Environ.*, **257**, 121-132.

Tornaghi, M. E., Premoli Silva, I. and Ripepe, M. 1989, Lithostratigraphy and planktonic foraminiferal biostratigraphy of the Aptian-Albian "Scisti a Fucoidi" in the Piobbicocore, Marche, Italy: background for cyclostratigraphy. *Riv. It. Paleontol. Strat.*, **95**, 223-264.

Trask, P. D. (ed.)(1939), *Recent Marine Sediments*. Am. Assoc. Petrol. Geol.

Tsuji, T. and Miyata, Y. (1987), Fluidization and liquefaction of sand beds: Experimental study and examples from Nichinan Group. *J. Geol. Soc. Japan*, **93** (11), 791-808. (J+E)

Tucker, M. (2002), Fifty years of the IAS and *Sedimentology. IAS Newsletter*, (178), 3-4.

Twenhofel, W. H. (1932), *Treatise on Sedimentation. 2nd Ed.* Williams & Wilkins, Baltimore.

_____ (1941), The frontiers of sedimentary mineralogy and petrology. *J. Sed. Petrol.*, **11** (2), 53-63.

_____ (1950), *Principles of Sedimentation. 2nd Ed.* McGraw-Hill, New York.

_____ (1961), *Treatise on Sedimentation. New Dover Edition. Vol. 1.* Dover Publications, Inc., New York.

Tyrrell, G. W. (1933), Greenstones and greywackes. *Reun. Intern. Pour l'Etude du Précamb. et des Vieilles Chaines de Montagnes en Finlande, Comp. Rend.*, 1931, 24-26.

Udden, J. A. (1914), Mechanical composition of clastic sediments. *Geol. Soc. Am. Bull.*, **25**, 655-744.

Uemura, T. (1981), Deformation facies, series and grades. *J. Geol. Soc. Japan*, **87** (5), 297-305.

Ujiié, H., Nakamura, T., Miyamoto, Y., Park, J.-O., Hyun, S. and Oyakawa, T. (1997), Holocene turbidite cores from the southern Ryukyu Trench slope: suggestions of periodic earthquakes. *J. Geol. Soc. Japan*, **103** (6), 590-603.

Ujiie, Y. (1981), Diagenetic alteration of kerogen in the Cretaceous and Tertiary Systems of the Tempoku district, Hokkaido, Japan. *J. Geol. Soc. Japan*, **87** (4), 225-237. (J+E)

Ussher, J. (1650-1654), *Annales Veteris et Novi Testamenti* (Part 1 1650; Part 2 1654). Flesher, London.

Utada, M. (1966), Zonal distribution of authigenic zeolites in the Tertiary pyroclastic rocks in Mogami district, Yamagata Prefecture. *Sci. Paper, Coll. Gen. Educ. Univ. Tokyo*, **15**, 173-216.

Uyeda, S. and Miyashiro, A. (1974), Plate tectonics and the Japanese Islands: a synthesis. *Geol. Soc. Am. Bull.*, **85**, 1159-1170.

Vail, P. R. (1987), Seismic stratigraphy interpretation using sequence stratigraphy. Part 1. Seismic stratigraphy interpretation procedure. *In*: Bally, A.S. (ed.), *Atlas of Seismic Stratigraphy. Vol. 1. Am. Assoc. Petrol. Geol., Studies in Geology*, **27**, 1-10.

_____ Mitchum, R. M., Jr. and Thompson, S., III (1977a), Seismic stratigraphy and global changes of sea level. Part 3: Relative changes of sea level from coastal onlap. *In*: Payton, C. E. (ed.), *Seismic Stratigraphy – Application to Hydrocarbon Exploration. Am. Assoc. Petrol. Geol. Mem.*, **26**, 63-81.

_____ _____ _____ III (1977b), Seismic stratigraphy and global changes of sea level. Part 4: Global cycles of relative changes of sea level. *In*: Payton, C. E. (ed.), *Seismic Stratigraphy – Application to Hydrocarbon Exploration. Am. Assoc. Petrol. Geol. Mem.*, **26**, 83-97.

Vallier, T., Thied, J. *et al.* (1979), In the Mid-Pacific Leg 62 probes the paleo-environments. *Geotimes*, **24** (2), 24-26.

van Straaten, L. M. J. U. (1954), Comparison and structure of recent marine sediments in the Netherlands. *Leid. Geol. Meded.*, **19**, 1-110.

_____ (1961), Sedimentation in tidal flat areas. *Alberta Soc. Petrol. Geol. J.*, **9**, 203-213, 216-226.

_____ (1964), *Deltaic and Shallow Marine Deposits. Developments in Sedimentology. 1*, Elsevier, New York.

van Watershoot van der Gracht, W. A. J. M. (1931), Permo-Carboniferous orogeny in south-central United States. *Am. Assoc. Petrol. Geol. Bull.*, **15**, 991-1057.
Vatan, A. (1967), *Manuel de Sédimentologie*. Technip, Paris.
Vening Meinesz, F. A. (1934), *Gravity Expeditions at Sea. V. 2.* Waltmann, Delft.
_____ (1940), The earth's crust deformation in the East Indies. *Kon. Ned. Akad. Wetenschap., Proc., Ser. B*, **43**, 278-306.
_____ (1955), Plastic buckling of the earth's crust: the origin of geosynclines. *In*: Poldervaart, A. (ed.), *Crust of the Earth*. Geol. Soc. Am. Spec. Paper, **62**, 319-330.
Vogt, T. (1930), On the chronological order of deposition of the Highland Schists. *Geol. Mag.*, **67**, 68-73.
Wadell, H. (1932a), Volume, shape and roundness of rock particles. *J. Geol.*, **40**, 443-451.
_____ (1932b), Sedimentation and sedimentology. *Science*, 75, 20.
_____ (1933a), Sedimentation and sedimentology. *Science*, 77, 536-537.
_____ (1933b), Sphericity and roundness of quartz particles. *J. Geol.*, **41**, 310-331.
Wakita, K. (1988a), Early Cretaceous mélange in the Hida-kanayama area, central Japan. *Bull. Geol. Surv. Japan*, **39**, 367-421.
_____ (1988b), Origin of chaotically mixed rock bodies in the Early Jurassic to Early Cretaceous sedimentary complex of the Mino terrane, central Japan. *Bull. Geol. Surv. Japan*, **39**, 675-757.
Walker, R. G. (1965), The origin and significance of the internal sedimentary structures of turbidites. *Proc. Yorkshire Geol. Soc.*, **35** (1), 1-32.
_____ (1967), Turbidite sedimentary structures and their relationship to proximal and distal depositional environments.. *J Sed. Petrol.*, **37** (1), 25-43.
_____ (1975), Generalized facies models for resedimented conglomerates of turbidite association. *Geol. Soc. Am. Bull.*, **86**, 737-748.
_____ (1978), Deep-water sandstone facies and ancient submarine fans: models for exploration for stratigraphic traps. *Am. Assoc. Petrol. Geol. Bull.*, **62**, 932-966.
_____ (1979), Facies and facies models. General introduction. *In*: Walker, R. G. (ed.), *Facies Models*. Geoscience Canada, Reprint Series 1, Geol. Soc. Canada, 1-8.
_____ and James, N. P. (eds.)(1992), *Facies Models. Response to Sea Level Change*. Geol. Assoc. Canada, St. John's.
Walther, J. (1893), *Einleitung in die Geologie als Historische Wissenschaft. Bd. 1. Beobachtungen über die Bildung der Gestine und ihrer organischen Einschlüsse.* Gustav Fischer, Jena.
_____ (1894), *Einleitung in die Geologie als Historische Wissenschaft. Bd. 3. Lithogenesis der Gegenwart.* Gustav Fischer, Jena, 535-1055.
Watanabe, H. (ed.) (1935), *Dictionary of Science of the Earth*. Kokon Shoin Co. Ltd., Tokyo. (J)
Weatherly, G. L. and Kelley, E. A., Jr. (1982), 'Too cold' bottom layers at the base of the Scotian Rise. *J. Mar. Res.*, **40** (4), 985-1012.
Wentworth, C. K. (1919), A laboratory and field study of cobble abrasion.*J. Geol.*, **27**, 507-521.
_____ (1922), A scale of grade and class terms for clastic sediments. *J. Geol.*, **30**, 377-392.
Werner, A. G. (1786), *Short Classification and Description of the Various Rocks*. Ospovat, A. (trans. 1971). Hafner, New York.
Whewell, A. (1832), Review of Lyell, 1830-3. vol. ii. *Quarterly Review*, **47**, 103-132.
Whitaker, J. H. McD. (1974), Ancient submarine canyons and fan valleys. *Soc. Econ. Paleont. Mineral., Spec. Publ.*, **19**, 106-125.
_____ (1976*), Submarine Canyons and Deep-Sea Fans, Modern and Ancient.* Dowden, Hutchinson & Ross, Stroudsburg, Pa.
Wilcockson, W. H. (1945), The geological work of Henry Clifton Sorby. *Proc. Yorkshire Geol. Soc.*, **27**, 1-22.
Williams, D. F., Lerche, I. and Full, W. E. (1988), *Isotope Chronostratigraphy: Theory and Methods*. Academic Press, San Diego.

Williams, E. G. (1965), Memorial to Paul D. Krynine (1901-1964). *Geol. Soc. Am., Proc.* (1965), 63-67.
Williams, G. P. (1970), Flume width and water depth effects in sediment transport experiments. *U.S. Geol. Surv. Prof. Paper*, **562-H**, 1-37.
Wilson, C. (ed.)(1994), *Earth Heritage Conservation.* The Geological Society/The Open University.
Wilson, J. L. (1975), *Carbonate Facies in Geologic History.* Springer-Verlag, New York.
Wilson, R. C. L. (1991), Sequence stratigraphy: an introduction. *Geoscientist*, **1** (1), 13-23.
_____ (1998), Sequence stratigraphy: a revolution without a cause? *In*: Blundell, D. J. and Scott, A. C. (eds.), *Lyell: the Past is the Key to the Present. Geol. Soc. London, Spec. Publ.*, **143**, 303-314.
Wood, A. and Smith, A. J. (1959), The sedimentation and sedimentary history of the Aberystwyth Grits (Upper Llandoverian). *Quart. J. Geol. Soc. London*, **114**, 163-195.
Woodward, H. B. (1911), *History of Geology.* Watts & Co., London.
Wüst, G. (1936), Schichtung und Zirkulation des Atlantischen Ozeans. Das Bodenwasser und die Stratosphäre. *Wiss. Erg. Dtsch. Atlant. "Meteor" 1925-27*, **6**, 1-288.
Yabe, H. (1927), Cretaceous stratigraphy of the Japanese Islands. *Sci. Repts. Tohoku Univ., 2nd Ser. (Geol.)*, **11** (1), 27-100.
Yagi, T. (1929), Chemical features of marine shales in Hokkaido and characteristics of their submarine weathering. *J. Japan. Assoc. Min. Petr. Econ. Geol.*, **1** (2), 17-29. (J)
_____ (1930), Glauconites in Japan (1)(2). *J. Japan. Assoc. Min. Petr. Econ. Geol.*,, 3 (3), 119-129; 3 (4), 174-180. (J)
_____ (1933), Sedimentological study of the lower formation in the Tsugaru and Matsumae oil fields (1). *J. Japan. Assoc. Min. Petr. Econ. Geol.*, **10** (3), 82-92; (4), 116-123. (J)
_____ (1935), Sedimentological study of the lower formation in the Tsugaru and Matsumae oil fields (2). *J. Japan. Assoc. Min. Petr. Econ. Geol.*, **13** (1), 17-24; (2), 52-58. (J)
Yagishita, K. (2001), *Facies Analysis and Sedimentary Structures.* Kokon Shoin Co. Ltd., Tokyo. (J)
_____ and Taira, A. (1989), Grain fabric of a laboratory antidune. *Sedimentology*, **36**, 1001-1005.
_____ Arakawa, S. and Taira, A. (1992), Grain fabric of hummocky and swaley cross-stratification. *Sed. Geol.*, **78**, 181-189.
Yamada, T. (2003), Stenonian revolution or Leibnizian revival?: Constructing geo-history in the seventeenth century. *Histora Sceientiarum*, **13-2**, 75-100.
Yamaguchi, K. (1939), Memorial to Hanjiro Imai, Dr. Sci. *J. Geol. Soc. Japan*, **46** (545). (J)
Yamamoto, K. (1983), Geochemical study of Triassic bedded cherts from Kamiaso, Gifu Prefecture. *J. Geol. Soc. Japan*, **89** (3), 143-162. (J+E)
Yamamoto, S. (1988), Thickness distribution of pelagic clay and bedded chert in the North Pacific. *J. Sed. Soc. Japan*, (28), 27-39. (J+E)
_____ (1989), Stratigraphy and genesis of cherts in Chirai Quarry, Nikoro Group, Hokkaido: Comparison with stratigraphy of the DSDP site. *J. Sed. Soc. Japan*, (31), 15-24. (J+E)
Yamauchi, S. (1977), On the slump structures in the Miocene series of the Chichibu basin, central Japan. Part 1, Morphology. *J. Geol. Soc. Japan*, **83** (8), 475-489. (J+E)
Yatsu, E. (1954–55), On the longitudinal profiles of the graded river. *Miscellaneous Reports of the Research Institute for Natural Resources*, (33), 15-24; (34), 14-21; (35), 1-6. (J+E)
_____ (1955), On the formation of the slope discontinuity at the fan margins. *Misc. Rep. Res. Inst. Natur. Resources*, (36), 57-64.
Yeakel, L. S., Jr. (1962), Tuscarora, Juniata, and Bald Eagle paleocurrents and paleogeography in the Central Appalachians. *Geol. Soc. Am. Bull.*, **73**, 1515-1540.
Yokokawa, M. (1995), Combined-flow ripples: genetic experiments and applications for geologic records. *Mem. Fac. Sci., Kyushu Univ., Ser. D*, **29** (1), 1-38.
_____ Masuda, F., Sakai, T., Endo, N. and Kubo, Y. (1999), Sedimentary structures generated

in upper-flow-regime with sediment supply: Antidune cross-stratification (HCS mimics) in a flume. *In*: Saito, Y. *et al.* (eds.), *Land-Sea Link in Asia*. STA(JISTEC) & Geol. Surv. Japan, 409-414.

Yokoyama, M. (1914), *Futsu Chishitsu-gaku Kogi (Lectures of General Geology).* Fuzanbo, Tokyo. (J)

Yoshimura, S. (1937), *Kosho-gaku (Limnology).* Sanseido, Tokyo. (J)

_____ (1976), *Kosho-gaku (Limnology).* Enlarged edition. Seisan Gijutsu Center, Tokyo. (J)

Young, R. N. and Southard, J. B. (1978), Erosion of fine-grained marine sediments: sea-floor and laboratory experiments. *Geol. Soc. Am. Bull.*, **89**, 663-672.

Ziegler, A. M. (1965), Silurian marine communities and their environmental significance. *Nature*, **207**, 270-272.

Zimmerle, W. (1968), Serpentine graywackes from the North Coast basin, Colombia, and their geotectonic significance. *Neues Jahrb. Mineral. Abh.*, **109**, 156-182.

Zingg, T. (1935), Beiträge zur Schotteranalyse, Schweiz. *Min. Petrog. Mitt.*, **15**, 39-140.

Zuffa, G. G. (1985), Optical analyses of arenites: influence of methodology on compositional results. *In*: Zuffa, G. G. (ed.), *Provenance of Arenites*. D. Reidel Publ., Dordrecht, 165-189.

# Index of Names

Figures are indicated by page number entries in *italics*

# Index of Subjects

Figures are indicated by page number entries in *italics*